T0083612

SIGILLVM · VNIVERSITATIS · SCOLARIVM · STVDII · PRAGENSIS

Irena Štěpánová

Newton
Kosmos – Bios – Logos

CHARLES UNIVERSITY IN PRAGUE
KAROLINUM PRESS 2014

Reviewed by: Stephen David Snobelen, Associate Professor, History of Science and Technology Programme, Kings College, University of Halifax, Canada. Director of Newton Project Canada

CATALOGUING-IN-PUBLICATION – NATIONAL LIBRARY OF THE CZECH REPUBLIC

Štěpánová, Irena
Newton: Kosmos, Bios, Logos / Irena Štěpánová. – Prague : Karolinum Press, 2014
Edited by: Charles University in Prague
ISBN 978-80-246-2379-5

141.4 * 101 * 2-1 * 133 * 141.331 * 113/119 * (410.1)
- Newton, Isaac, Sir, 1643–1727
- 1643–1727
- God – philosophical approach
- God – esoteric approach
- theology – esoteric approach
- hermetism
- philosophy of nature – England – 17th–18th centuries
- monographs

210 – Philosophy and theory of religion [5]

ISBN 978-80-246-2379-5
ISBN 978-80-246-2389-4 (pdf)

Content

Acknowledgements

My greatest thanks go to Zdenek F. Danes, whose help with the translation of my original text was crucial.

And I would also like to appreciate the three ladies, whose help was great: editors Alena Jirsová and Martina Pranic, and my daughter Petra.

I. Introduction

Newton was not the first of the age of reason. He was the last of the magicians, the last of the Babylonians and Sumerians, the last great mind which looked out on the visible and intellectual world with the same eyes as those who began to build our intellectual inheritance rather less than 10 000 years ago. Isaac Newton [...] was the last wonderchild to whom the Magi could do sincere and appropriate homage.
John M. Keynes[1]

A belief that our comprehension of this world keeps increasing as if our heads were some endlessly inflatable balloons is common. However, the reality is different. Our knowledge resembles sedimentation: new information covers up old knowledge and pushes it into oblivion. While gaining new insights, we lose the wisdom of old. Certainly, some of that loss we may never regret: but the process of sedimentation may also obscure what should have been remembered. We thus may have lost a part of ourselves.

Fortunately, from time to time, and often after years of concentrated effort, we happily return to long-forgotten, even rejected, knowledge. A case in point is *hermetic philosophy*: not just as an example of "recurring" knowledge, but also as a record of gradual change of the overall frame of our learning, of our method, and, eventually, of our way of thinking. Hermetic philosophy (and alchemy as its practical part) represents an entirely different relation to natural world from what corresponds to our abstract rational approach. In fact, it seems to be an ideal topic to study the "history of ideas."

1 John Maynard Keynes: "Newton, the man," in *Essays and Sketches in Biography*, New York: Meridian Books 1956, p. 280.

The present book is a brief effort to show whether, and to what extent, hermetic philosophy may have inspired one of the founders of modern European science.

II. Sources of Newton's Inspiration

Nemo suscipiet caelum; religiosus pro insano, inreligiosus putabitur prudens, furiosus fortis, pro bono habebitur pessimus [...] Haec et talis senectus ueniet mundi: irreligio, inordinatio, inrationabilitas bonorum omnium.
Asclepius, c. 2nd century.[2]

Wisdom has irretrievably succumbed to news reporting, shallow entertainment and demand. While the past all was (allegedly) rational and serious, now we are prisoners of reports. They float like dust and make existential appropriation of being – as the philosophers call it – impossible.
Petra Gümplová, 2007.[3]

In every age there were people who clearly saw that in the course of time the human spiritual level changes in a strange way: while knowledge naturally increases, spiritually mankind sinks ever lower. More than eighteen centuries separate the two quotations presented as the central pieces of this chapter, yet both say the same: in earlier times mankind was, spiritually, better off. It was closer to the mystical Beginning. And this idea, too, is characteristic for Isaac Newton: it haunted him.

For some time, it has been clear and generally accepted that Newton believed in *prisca sapientia,*[4] that he frequently quoted authors from

2 Asclepius, verses 25–26. In: A.-J. Festugiére (ed.): *Corpus hermeticum, Tome 2 – Traités 13–18, Asclepius*, Paris: Belles-Lettres, 1983, p. 329. "Nobody will look up to heavens. Religious will be called insane, irreligious prudent, furious strong, the worst one will be called good... This is the world's senility: lack of religion, lack of order, lack of all reasonable goods."

3 Petra Gümplová: "Ztraceno v blábolu," in: *Pátek Lidových novin*, 16. 3. 2007.

4 E.g., Steven D. Snobelen: "God of Gods, and Lord of Lords: the theology of Isaac Newton's General Scholium to the Principia," *Osiris* 16, pp. 169–208, here p. 185.

antiquity and that he – so to speak – felt that he was continuing in the antique priest-scientist tradition.[5]

We shall see that he saw himself more as a person who revives the half-forgotten antique wisdom than as a discoverer of entirely new ways of thinking. However, it has not yet occurred to anyone that Newton and his intellectual world literally derived from antiquity. Perhaps no one has yet seriously considered the possibility that the father of European science could have bypassed centuries of evolving European ideas and resumed an ancient line of thought. Nevertheless, we shall try to prove that modern science owes its beginning to Newton's precise following of some thinking patterns that date precisely from ancient times.

Newton's inspirations have been thoroughly studied by a number of authors; e.g. the prominent American scholar, Betty Jo Teeter Dobbs, dedicates a substantial portion of her books on Newton-the-Alchemist to this very problem.[6]

For our purposes, we shall use those sources that may enrich present scholarship in Newtonian studies and open up new topics in them.

The Hexameral literature and the Bible

By Hexameral literature[7] we mean those texts that study the six days of creation according to the First Book of Moses, Genesis, Ch. 1, verses 1–27.[8] Although that type of literature is of a very ancient date, starting with Origen around the middle of the 3rd century and ending with John Milton in the 17th century, in Newton's time it was still a matter of interest.

The creation of the world according to the Bible is a mythical event, and, as such, has a timeless meaning: being a myth, it gives man a chance to think about himself and about his place in the world. Newton was certainly one of those who were fully aware that they have to deal with a truth of a higher order, which not only agrees with reality but also raises moral demands, and wields a great power, because it reaches

5 Ibid., p. 187.

6 Betty Jo Teeter Dobbs: *The Foundations of Newton's Alchemy*, Cambridge: Cambridge University Press, 1975. Betty Jo Teeter Dobbs: *The Janus Faces of Genius*, Cambridge: Cambridge University Press, 2002.

7 From Greek ἕζ – six, and ἡμέρα – day.

8 We use here the King James Bible.

beyond plain reason. Myth reaches all the way into the realm of values and emotions.[9]

In the 14th century, Henry von Langenstein wrote an influential book, *Lecturae super Genesim*,[10] where he quotes sixty-four authors and their explanations of the creation of the world; the authors are not only Christian, but also pre-Christian, Arabic, Greek, Roman and Jewish writers. The Hexameral commentaries may be understood as the focus of the beginning of European science. Those studies always tried to find a common ground between Moses' mythical concept of creation and the results of natural philosophy. Perhaps, with some exaggeration, we may claim that natural science gradually arose in the emancipation of Hexameral authors from the confines of Biblical exegesis.

Newton possessed a thorough knowledge of the Bible, and there is no doubt that, due to his profound religiousness, it was an important source of inspiration throughout his life. Here is an instance of Newton's Hexameral commentary touching upon the actual duration of those six days of creation. Newton's acuity is conspicuous:

> You may make ye first day as long as you please & ye second day too if there was no diurnal motion till there was a terraqueous globe, that is till towards ye end of that days work.[11]

We think, moreover, that Hexameral literature influenced Newton's methodology: his division of the world is based on Biblical Genesis, 1,1–27.

The text describes the creation of the world in three steps. Three times the text uses the Hebrew word *bara* which, in the Old Testament, is exclusively reserved for Divine activity. We translate it as "create," however, the Hebrew original has a profound meaning which we no longer recognize at the present time. Creation in the Hebrew meaning is far beyond human capability. Man always makes one thing out of another: divine Creation is something out of nothing. Not only that, God always made something absolutely new, something that existed never before and did not follow from anything that had been created earlier. Triple use of the word *bara* means that the world was made in three steps, the later and higher levels always being something absolutely, revolutionary new.

---

9 Jan Assmann: *Kultura a paměť* (i.e. *Culture and Memory*), Praha: Prostor 2001, p. 70.

10 Betty Jo Teeter Dobbs: *The Janus Faces of Genius*, p. 58.

11 Betty Jo Teeter Dobbs: *The Janus Faces of Genius*, p. 62.

We thus reason that the division according to Genesis influenced the Newton's methodological thinking and division of his work. We try to show it in the following table. Corresponding Biblical verses are in the left column. It helps to read the table from the left lower corner and read upward and toward the right side.[12]

3.	*Gen 1,27: So God created man in his own image, in the image of God created he him; male and female created he them.*	***LOGOS*** *Domain of meaning. Man as God's image + Divine Providence.*	***Freedom*** *of law and determination.*	***History and theology*** *as study of man's action and God's providence. (Nowadays Humanities, but with a major drawback: man is not an object.)*
2.	*Gen 1,21: And God created great whales, and every living creature that moveth, ...*	***BIOS*** *Life is implanted into matter.*	***Determination*** *often paradoxical with respect to level 1.*	***Alchemy*** *(nowadays Biology, but, so far, does not know what life is.)*
1.	*Gen 1,1: In the beginning God created the heaven and the earth.*	***KOSMOS*** *Lifeless matter.*	***Natural laws*** *are valid.*	***Natural philosophy*** *(nowadays Natural Sciences).*

The First Domain (starting from the bottom) is Nature without life. God is understood in His intelligent plan that seems to require an Intelligent Creator (the so-called *Design Argument*, see chapter V).

The Second Domain is the domain of Life. Since times immemorial, it was the subject matter of alchemy, which profoundly occupied Newton for a long time. We shall return to it in Chapter IV.

The Third Domain is the most mysterious. For the time being, we shall call it the Domain of the Logos.

The Greek word λογοσ (logos) has several meanings: word, language, even idea.[13] Originally, it meant a collection, an assembly of items that naturally fit together.

---

12 Our inspiration came from two publications: Zdeněk Trtík: *Vztah já-ty a křesťanství*, (ie. *Relation Me-You and Christianity*) Praha: Ústřední rada ČSH, 1948, and Zdeněk Neubauer: "Apotheosy of Metamorphosis," in *Akademie u sv. Mikuláše, Anthology 2004/2005*, Praha: Blahoslav, 2005.

13 Berry, George Ricker: *The Classic Greek Dictionary*, Follett Publishing Co., New York, Chicago, Pasadena, 1958:
ὁ λόγος word, language, talk, pretence, saying, expression, oracle, maxim, proverb, conversation, discussion, conference, interview, speaking, talking, rumor, tale, story, fable, narrative, history,

Gradually, its meaning was reduced to linguistic usage and it best fits our word "meaning."

"Meaning" points toward "togetherness," it is an interconnection of what meaningfully belongs together. We can express such a meaning only by means of language (again *logos*), a unique possession of man as God's image.

Human language is the only means of comprehending the world and pointing toward its meaning. And language is in fact the only way to carry out this comprehending: it creates a web connecting all those individual events.[14] Those events make up the essence of the world's history.

We believe that Newton understood the Third Domain as the domain of history, where God and man cooperate as active partners. That is also a heritage of the Old Testament: history is a realm both human and divine. God and man work together in making history. More on that matter will follow in Chapter III.

Newton was not only a modern scientist: he also enjoyed solving the riddles so very popular in the Renaissance. We believe that it was the mystical event from Biblical Genesis and its commentary that directed his methodological conclusions in that field.

Philo of Alexandria (15 B.C. - A.D. 50)

Philo of Alexandria was a Greco-Jewish philosopher educated in the tradition of the Book of Wisdom. He was well-versed in the Old Testament as well as in Poseidonius, and made full use of that knowledge in his work.[15] Today, he represents the mid-Platonic philosophy. Philo tried to join two mutually exclusive domains – philosophy and faith. This alone interested Newton[16] who, likewise, tried to combine the opposites in se-

chronicle, fable, prose, book, speech, eloquence, account, consideration, esteem, regard, calculation, reckoning, relation, proportion, analogy, condition, reason.
In the New Testament: Λόγος Jesus Christ.

14 Zdeněk Neubauer: "Do světa na zkušenou" (ie. an essay about Tolkien's work), in: *Dodatky k Silmarillionu, Studijní materiál pro potřeby Tolkienovského semináře při Parconu*, ed. Michal Bronec, 1990, p. 39. Furthermore Gregory Bateson: *Mind and Nature: A necessary unity*, Cresskill, N.Y.: Hampton Press, 2002.

15 Zdeněk Kratochvíl: *Prolínání světů*, Praha: Herrmann a synové, 1991, p. 15.

16 According to the book by John Harrison: *The library of Isaac Newton*, Cambridge and New York: Cambridge University Press, 1978, p. 216, Newton had in his library a book by this author: 1300 PHILO, Judaeus: *Omnia quæ extant opera. Ex accuratissima S. Gelenii, & aliorum interpretatione... (Greek & Latin)*, Lutetiæ Parisiorum, 1640.

veral disciplines. We are interested in his work, too, when we investigate the influence of emotional matters upon strictly rational thinking.

Such a joining of wide-ranging influences, typical for Philo's times, is what we now call *syncretism*. Philo was the first who tried to transform the Hebrew legacy into a new doctrine similar to Greek philosophy. He is today known as an inventor of new method, called allegorical exegesis. Philo felt the pressure that the modern man knows quite well: how to retain one's piety when rational criticism threatens the meaning of a sacred text.

Philo interpreted the texts allegorically in order to express their spiritual message. He tried to extend their meaning to encompass the wholeness of the world by means of interpretation, which in fact made up the translation between two cultural areas, Hebrew and Hellenic. It requires a conscious categorization of events into principles, and can be done only at the philosophical level of thinking.[17]

Although Philo tried to see God as a living entity, close to the Stoic interpretation, he simultaneously shared the Platonic resistance toward everything material.

Thus God fills everything and encompasses everything in His vital activity, yet He Himself cannot be comprehended: He is One and Everything (*heis kai to pan*; with the Neo-Platonists that term is transformed back into the neuter *to hen kai pan*).[18]

Philo finds that matter is the ultimate evil. Therefore his concept of God is purely transcendental.[19] Although Philo had a considerable influence upon Newton, in this fundamental respect Newton departed from him. This will be shown in the analysis of Scholium generale in Chapter VI.

Philo is also connected with the origins of the Alexandrean Metaphysics of the Logos, which, unlike the classical metaphysics, is dynamic. It is therefore questionable whether it is metaphysics at all.

As a rule, European metaphysics studies unchangeable, transcendental principles beyond experience, and examines rational cases. On the other hand, mid-Platonic Philo investigates existence and comprehensibility. Those depend on movement, not on immobility.[20]

---

17 Zdeněk Kratochvíl: *Prolínání světů*, p. 25.

18 Ibid., p. 28.

19 František Kovář: *Filosofické myšlení hellenistického židovstva*, Praha: Herrmann a synové, 1996, p. 183;
Ivo Tretera: *Nástin dějin evropského myšlení*, Litomyšl: Paseka 2002, p. 127.

20 Zdeněk Kratochvíl: *Prolínání světů*, p. 30.

Ancient Egypt

The Old Testament describes Egypt as a place of utmost decadence, idolatry, zoophilia, superstition and all kind of abomination that may be overcome by nothing less than exodus and, eventually, by complete oblivion. In other words: it demands an active removal of all reminiscences.

This attitude prevailed in Christianity until the Renaissance, when the opinion changed dramatically. Egypt became a source of everything worthy that came later. It then became the true beginning of the spiritual evolution which advanced via the exodus and Judaism and progressed toward Christ and Christianity. And the 17th century turned the ideas about ancient Egypt into a complete Egyptomania, one that reached its climax in the time of the Enlightenment.[21]

Of course, Christian scholars could not immediately study Egypt, since, for the orthodoxy, Egypt was still the hated paganism incarnate. Such scholars could be accused of heresy and persecuted. But those Biblical scholars who wanted to study secrets of ancient Egypt without prejudice found a way around, due to their thorough knowledge of the Scriptures.

Scholars of Newton's times protected themselves from possible persecution by a single verse from the New Testament, the Book of Acts of the Apostles 7,22.

Stephen the Martyr, in his farewell address before he was executed by stoning, said about Moses:

> And Moses was learned in all the wisdom of the Egyptians, and was mighty in words and in deeds.

In the entire Bible, that is the only favourable sentence about Egypt. In the sub-chapters about Spencer and Cudworth, we shall show how this single sentence opened the door for their unexpected and enormous intellectual achievements.

Maimonides (1135–1204)

The Jewish scholar Maimonides[22] (Rabi Moses ben Maimon) was the supreme authority for the Protestant scholars of the 17th century. With

21 Jan Assmann: *Moses the Egyptian*, Frankfurt am Main: Fischer Taschenbuch Verlag 2004, p. 85 and afterwards.

22 Ibid., pp. 88–92.

his ideas, more than five centuries old, he created a way out for their further studies.

At Cambridge, his ideas were taught by two prominent scholars:[23] John Spencer (1630–1693), an expert on Hebrew, and Edward Pococke (1604–1692), expert on Hebrew and Arabic.[24]

In his library, Newton had several books by Maimonides,[25] so that he had first-hand access to his ideas. Maimonides was also Newton's main source for his study of Jewish history.[26] Besides, he was indirectly exposed to his ideas, as will be shown in the next chapter.

Like the aforementioned Philo, Maimonides had to deal with the conflict between rationality and sacred Jewish texts. He came up with the idea that every one of the 613 commandments of the Torah must have had some rational purpose; and if we cannot find it, we have to look for a historical explanation. According to Maimonides, God creates human history in the same way as He created nature. There are no sudden breaks. Everything follows organically from what preceded: *Natura non facit saltus* – Nature does not take leaps. From one extreme to another, things progress through a series of infinitesimal small steps. Incidentally, this idea may have inspired Newton for his infinitesimal calculus as a mathematical method of studying infinitely small quantities and their changes.

According to Maimonides, divine wisdom is revealed in those continuous and infinitely small movements, gradual changes and natural growth. This was a very serious step: it was contrary to the traditional division of natural and revealed religion. According to Maimonides, even

23 Frank E. Manuel: *The Religion of Isaac Newton*, p. 66.

24 Edward Pocock is referred to by Newton as "our Pocock" right in his *Scholium Generale*. According to the book that gives a complete catalogue of Newton's library:
John Harrison: *The library of Isaac Newton*, Cambridge and New York: Cambridge University Press, 1978, p. 219, Newton had Pococke's book that connects with Maimonides:
Pococke, Edward: Porta Mosis, sive dissertatioaliquot a R. Mose Maimonide (2 pts.), 1655.

25 Here we list the titles and catalogue numbers of books according to: John Harrison: *The library of Isaac Newton*, p. 186:
1018 Maimonides: *De cultu divino ex R. Mosis Majemonidæ secunda lege, seu Manu forti liber VIII*, Parisiis, 1687.
1019 Maimonides: *De idolatria liber, cum interpretatione Latina & notis D. Vossi*, Amsterdami, 1641.
1020 Maimonides: *De sacrificiis liber. Accesserunt Abarbanelis Exordium.* Londini, 1683.
1021 Maimonides: *Porta Mosis, sive Dissertationes aliquot...Nunc primum arabice... &Latine editæ... Opera & studio E. Pocockii*, Oxoniæ, 1655.
1022 Maimonides: *Tractatus de iuribus anni septimi et iubulaei*, Francofurti ad Meonum, 1708.

26 Matt Goldish: Newton on Kabbalah, in *The book of nature and scripture, recent essays on natural philosophy, theology, and Biblical criticism*, eds. James E. Force – Richard H. Popkin, Dordrecht: Kluwer Academic Publishers, 1994, p. 90.

revelation is a natural historical process; and, likewise, natural phenomena are guided by God's hand.

Maimonides was inspired by Manetho[27] and followed his concept of "normative inversion." To this day, Manetho is one of our sources of information about ancient Egypt. He came up with the idea of a "counter-community," an organized group of people who turn everything upside down. Whatever is mandatory in the original society, will be forbidden. And vice versa, whatever was forbidden, will be mandatory. This will in turn create something akin to "counter-laws."[28] Maimonides was not yet a historian in our sense of the word. His deductions may be called something like "historical apologetic theology." Maimonides started from the known history of the Jews and literally invented, according to Manetho's paradigm, their counter-society that worshipped everything that the Jews deplored, and forbade whatever the Jewish laws commanded.

Maimonides thus created the "Sabians," a fictitious nation, shaped perhaps after the Persians. Nevertheless, Maimonides herewith prepared the field for later actual historians, who replaced the fictitious Sabians, spiritual opponents of the Jews, by a real nation, the Egyptians.

The Cambridge Platonists

The Cambridge Platonists were a group of professors at Cambridge University toward the end of the 17th century. Their philosophy may be approximately defined as a combination of Neo-Platonism with stoicism and other influences. Their predecessors are mainly Philo of Alexandria and Justus Lipsius. One of the principal ideas of the Stoics, *pneuma*, a fine, fiery, all-penetrating substance, later refined into the neo-platonic non-material *aether*, was a concept that, for a long time, also interested Newton. It opened for him a new approach to the concept of force. Unlike his contemporary mechanists, for whom all forces acted by direct physical contacts, Newton, by means of this aether, could explain actions at a distance.

27 Manetho was an Egyptian priest who, in the mid-3rd century B.C., composed a book on the history of Egypt for the ruler, Ptolemaios II.

28 It would take us too far from our subject if we were to elaborate on the conditions that Manetho applied in his theory of counter-society and counter-laws: briefly, he tried to analyze the great trauma of Egyptian history, i.e. Akhaneton's forceful and unsuccessful experiment of a religious reform. Compare chapter Corpus hermeticum II. Jan Assmann: *Moses the Egyptian*, p. 55 et al.

Cambridge Platonists influenced not only Newton as a philosopher and physicist, but also Newton as a historian. It even impressed his concept of God.

Henry More (1614-1687)

Henry More is a more mystical, in fact theosophical, Cambridge Platonist. Later in life, he completely abandoned his strict Calvinist upbringing and devoted his life to the study of philosophy, mainly Neo-Platonic. It became his interest and perpetual joy for the rest of his life. More was highly productive and the brilliance of some of his early writings delights us to this day. We shall follow some of his ideas in Chapter V.

For instance: he used the term "spissitude," density, to describe the power of the spiritual realm in a particular place. Similar to the dimensions of a body, he used the terms "ana/kata". [29]

Newton owned several of his books,[30] one with a dedication written in the author's hand.

John Spencer (1630-1693)

John Spencer followed the ideas of Maimonides, but studied real history and found that the principle of *normative inversion* was hidden between

---

29 From Greek ανα - up, and κατα - from above, down, inside. These Greek prepositions are to this day used in technical terminology, although with a meaning different from More's: e.g., in medicine: *katabolism*.

30 According to: John Harrison: *The library of Isaac Newton*, pp. 195–6, Newton had several books by this author in his library, with the following titles and catalog numbers:
1110 More, Henry: *An antidote against atheisme,* London, 1653.
1111 More, Henry: *Apocalypsis Apocalypseos: or the Revelation of St. John unveiled*, London: 1680.
1112 More, Henry: *Discourses on several texts of scripture*, London: 1692.
1113 More, Henry: *The immortality of the Soul, so farre forth as it is demonstrable from the knowledge of nature and the light of reason*, London, 1659.
– *Observations upon Antroposophia theomagica, and Anima magica abscondita*, by Alazonomastix Philalethes [i.e. H. More], 1650, see 1199.
– *Paralipomena prophetica, containing several supplement and defences of Dr. Henry More*, 1685.
1114 More, Henry: *Philosophical poems*, Cambridge, 1647.
1115 More, Henry: *A plain and continued exposition of the several prophecies or divine visions of the Prophet Daniel*, London, 1681.
– *Remarks on Dr. Henry More's Expositions of the Apocalypse and Daniel*, 1690, see 1391.
1116 More, Henry: *Tetractys anti-astrologica, or, The four chapters in the Explanation of the grand mystery of Godliness*, London, 1681. With the author's dedication: *"Isaac Newton Donum Revern-dissimi Auctoris."*

the Jews and the Egyptians. He interpreted ingeniously the Sabians of Maimonides as pagans in general.[31]

He maintained a life-long interest in the ritual laws of the ancient Jews. It resulted in his book *De Legibus Hebraeorum Ritualibus et Earum Rationibus*. Newton owned a copy.[32] He transferred the interest from the conflict of Judaism and Christianity to an older one of Israel and Egypt.

Thanks to Spencer, pagan religion became for the first time a subject of serious scholarly study. His book on Egyptian rites was one of the sources of the Egyptomania of the 17th century. Spencer's actual knowledge of ancient Egypt became the precursors of modern religious studies and Egyptology despite the fact that he had to gather his information from Greek and Hellenic authors only. Spencer was a magnificent innovator. Unlike the earlier apologetic historical theology of Maimonides, Augustine, and Thomas Aquinas, Spencer's is a real historical research. He compared the canonical tradition with archaeological, epigraphic (non-canonical) and linguistic discoveries.

Spencer managed to show – and he did not deny that he owes it to the thoughts of Maimonides – that the Jewish ritual laws are evidently based on a denial of a previous, older religion, i.e. the religion of ancient Egypt; and that they arise from those antagonistic forces and negative potential of two *counter-religions*. According to that idea, Moses did not create his laws out of thin air: he just transformed the original idolatrous commandments.

Spencer made use of Maimonides' discovery that what we cannot achieve individually, we may accomplish as a society: we are capable of actively erasing *collective memory*. It was a great accomplishment. It meant that the only way to erase idolatry was a commandment of an exactly opposite rite to the common one.

Maimonides applied Manetho's original idea of *counter-community* in the sphere of religion. And here Maimonides applied his construct of *God's cunning*: The Jewish God dissolved the old rites by prescribing new, entirely opposite ones. That is the *normative inversion* of Maimonides. An example of this *inversion* is mentioned as far back as in Tacitus: the Jews sacrificed a lamb in fact to ridicule the supreme Egyptian god Amun, because the lamb was his sacred animal. Of course, God's *cunning* was

31 Jan Assmann: *Moses the Egyptian*, pp. 92–117.

32 In the catalog of his books: John Harrison: *The library of Isaac Newton*, p. 242: 1545 Spencer, John: *De Legibus Hebraeorum Ritualibus et Earum Rationibus*, Cantabrigiæ, 1685, p. 782.

successful: the Sabians of Maimonides were long-forgotten; and ancient Egypt, too, was condemned for a long, long oblivion.

The trick about the memory is as follows: in order to erase a memory, it has to be overwritten by an active *counter-memory*, ("ars oblivionis"). Such a *counter-memory* may be artificially constructed, but was justified and legalized as *God's intention*. The fact was that God was *cunning*, and that was the end of it.

Spencer surpassed Maimonides by reasoning several problems to the very end. For Maimonides, the Jewish law was timeless: once the law is valid, time is immaterial. On the other hand, Spencer saw the world through the eyes of a real historian: the law had its origin and, therefore, also an end. He was able to apply something that Maimonides was neither able, nor permitted to apply, i.e. the Christian evolutionism. When Spencer identified the origin of the laws of Torah as a defence against idolatry, he correctly reasoned that the Jews could have relaxed their rites when the danger to their true religion abetted and the rites lost their justification. According to the same reasoning, Spencer did not truly understand why the Jews had cursed our Saviour, who elevated love above the laws of Moses: the purpose of the law was after all to fight the danger of idolatry, and that danger was long gone. Spencer even brought adequate evidence that the true Egyptian zoolatry, at the time of Moses, was just an ancient history.

While Maimonides treated Revelation as a historical event and evidence of God's cunning, Spencer elaborated the detailed mechanism of the event.

He established not only the fact that Jewish laws were a transformation of Egyptian laws: he interpreted the Revelation with constructs *implantation*, *reception* and *transfer*. According to Spencer, Revelation is a gradual action. It is included in the process of gradual adaptation, change of explanation and change of codification.

Incidentally, Newton did not agree with Spencer's principal idea that Egypt was older than Israel. For him, Israel was the true original homeland, and the first revelation was given to Noah. For Newton, Egypt was just a decadence of the original revelation. It was only a temporary dwelling of the *chosen people* before their return to Palestine.

Spencer was only interested in rituals, not in theology; actual theology was studied in the same revolutionary manner by his colleague, Ralph Cudworth.

Ralph Cudworth (1617-1699)

Like John Spencer, Ralph Cudworth was one of the leading Hebraists at Cambridge University, and one of those scholars who were hiding from a potential persecution behind the aforementioned verse from the Acts of the Apostles. Thus they legitimized their historical research as theological scholarship, a search for Moses' original sources.

Cudworth's most important book is the voluminous *The True Intellectual System of the Universe.*[33]

Newton's *Out of Cudworth*,[34] i.e. handwritten notes on that book, shows how thoroughly Newton studied Cudworth's ideas.[35] We now have a transcript of those notes[36] they were published as an appendix to the book *Essays on the Context, Nature and Influence of Isaac Newton's Theology*,[37] we also have a brief description of the contents of those notes in the *Newtonproject,*[38] in addition to some remarks by authors who had the opportunity to read those notes.

For instance, Danton B. Sailor wrote that four sheets of the manuscript were full of stains and their corners were bent, so Newton must have studied Cudworth's entire *True Intellectual System*. He quotes a large number of passages verbatim, others he paraphrases, and only in a few places he adds his own commentaries.[39] We mention these details because Cudworth's book will be very important for our study of Newton's opus.

Cudworth had unusually thorough knowledge of ancient authors and made good use of it. His principal intention in his *True Intellectual System of the Universe* was to combat the growing atheism and, by means of arguments, show that atheism is an error. He tried to show that the very functioning of the cosmos proves that the cosmos had to be designed

33 Ralph Cudworth: *The True intellectual system of the Universe, wherein all the reason and Philosophy of Atheism is confuted, and its Impossibility Demonstrated*, London 1678. Later edition of this book, London: Thomas Tegg, 1845 – entire available online as a PDF file, see http://www.archive.org/details/trueintellectual03cudwuoft.

34 From Cudworth.

35 This book is not in the catalog of Newton's books; Newton owned another book, according to John Harrison: *The library of Isaac Newton*, p. 127.
466 Cudworth, Ralph: *A discourse concerning the true notion of the Lord's Supper*, London, 1670.

36 This manuscript of Newton is available at http://www.newtonproject.sussex.ac.uk/view/texts/normalized/THEM00118.

37 James E. Force and Richard H. Popkin (eds.): *Essays on the context, nature and influence of Isaac Newton's theology*, Dordrecht: Kluwer Academic, 1990.

38 www.newtonproject.sussex.ac.uk.

39 Danton B. Sailor: Newton's debt to Cudworth, *Journal of the History of Ideas* Nr. 49 (1988), pp. 511–518.

by an intelligent God (*design argument*) and that the atheists, by their opinion, reveal in fact their intellectual inferiority.

However, Cudworth employed his knowledge and highly rational argumentation for an end that appeared foolish for more than three centuries. He tried to disprove the findings of another highly educated scholar and linguist, Isaac Casaubon (1559–1614), on the authenticity of the *Corpus Hermeticum*. Cudworth believed that this document described the original, ancient Egyptian religion.

We shall focus on this problem and its solution, because it will help us to understand Newton's intellectual endeavour.[40]

Casaubon published his findings in 1614. He believed that his linguistic analysis definitely established that the texts of *Corpus Hermeticum* (*C.H.* hereafter) was written in *koine* Greek, a corrupt language; that they date from the 2nd to 3rd century; therefore it was impossible that they should authentically describe ancient Egyptian religion. Therefore, they were a fraud. According to Frances Yates, a recognized scholar, the year 1614 was a *watershed.* Assmann wrote:

> According to Frances Yates, the year 1614 in which [Casaubon's] book was published has to be recognized as a watershed separating the Renaissance world from the modern world ...[but]... Yates closed the book on the Hermetic tradition much too early. It was because of Cudworth's intervention and in Cudworth's interpretation that the Hermetic texts continued to be influential in the eighteenth century.[41]

Assmann further says that this question of authenticity of *C.H.* could not be adequately solved until recently, because in the 17th century nobody could read the hieroglyphs. Therefore, it was impossible to decide whether the religion in *C.H.* is described correctly or not. It took 200 years from the publication of Casaubon's book until 1822 when Jean-François Champollion deciphered the hieroglyphs, and another 200 years until the old Egyptian inscriptions were translated. Only at the present time can we decide with some confidence who was right; and the decision requires an Egyptologist. Therefore, only now can we also adequately judge Newton's conviction about *prisca sapientia* and about the question whether he could, or could not, be its archpriest, whether he stood in its line, which starts at the very dawn of cultural history.

---

40 Jan Assmann: *Moses the Egyptian*, pp. 118–130.

41 Jan Assmann: *Moses the Egyptian*, p. 85.

Cudworth was willing to agree with Casaubon that the texts of *C.H.* had been written in bad *koine* Greek. But he opposed him about their authenticity. He tried to show that at least some of them give a correct image of the ancient Egyptian religion. And the ancient Egyptian religion, similar to other religions of antiquity, was comprehended by Cudworth as follows: he used a large number of quotations from ancient authors to demonstrate that ancient civilizations worshiped many and various gods, but, at the same time, clearly recognized the existence of a supreme, non-created God.

Cudworth tried to reconstruct an old Egyptian theology in this way. That meant proving that besides the overt, exoteric religion for the common folk, who worshipped various gods and goddesses, the Egyptians knew a supreme God who did not have an exoteric cult; therefore, there existed an esoteric, secret theology for the initiated ones, an *arcane theology*.

He attempted to prove that the author of the hermetic texts, in his secret theology, was describing that One, Supreme and Universal God. This notion has not yet widespread: it is generally believed – undoubtedly under the influence of the Bible – that Egypt was the land of polytheism and that a true monotheism is tied to the Biblical religion only.

First, Cudworth presented a large anthology of Greek and Roman quotations from ancient authors, such as Origen, Climent of Alexandria, Plutarch, and several others, that clearly showed that ancient nations, besides an overt polytheism, did know a Supreme God.

He thus once and for all disproved the prejudice that it was only the Jews who in antiquity knew a Supreme God. Cudworth established that polytheism needs the existence of One, Supreme, Hidden God who penetrates all and holds everything together. He is the only one who can make this world function as a unity: one and everything, *hen kai pan*:

> First, the intelligent Pagans worshipped the one supreme God under many several names, secondly, that besides this one God they worshipped also many gods, that were indeed inferior deities subordinate to him, thirdly, that that they worshipped both the supreme and inferior gods, in images, statues and symbols, sometimes abusively called also gods.[42]

By means of many quotations, Cudworth was able to prove that ancient nations clearly distinguished between a supreme, non-created God

---

42 Ralph Cudworth: *The True Intellectual System of the Universe*, p. 197.

and a number of created gods; e.g., in Zoroastrism, Chaldeian religion and in Orphism, where we find the words *hen kai pan* for the first time.

For Cudworth, Egypt was the source of all wisdom. According to Egyptians, the world was a creation; not something that evolved spontaneously; that is documented by a series of Egyptian creation myths. There was thus a double Egyptian theology: one for the common people, vulgar, fabulous, and presented in events; and another, for the initiated, arcane and recognized, difficult, secret and hidden. And, for Cudworth, there were two ways to transfer the secret knowledge to future generations of the initiated ones: by allegorical myths and through the hieroglyphs.[43]

Cudworth searched all available antique documents and collected statements about a Supreme God. He found them in many authors. E.g., Plutarch mentions him in his treaty of Isis and Osiris. Horapollon calls God *Pantokrator* and *Kosmokrator*,[44] an omnipotent being that, from the unknown, governs the entire world.

Eusebius talked about a spiritual entity, full of reason and wisdom, the one that had created the world and is hard to find, because He is dark and hidden. Manetho wrote that in the popular theology god Amun gradually accepted the place of the highest god, because his name meant *hidden.*[45] According to Iamblichus, Amun is a demiurgic god and a representative of the highest truth. For Damascius, he is the principle, origin of all things and an invisible darkness.[46]

Cudworth saw ancient Egypt as follows: the supreme, hidden god, at first nameless, gradually fused with god Amun and thus received his exoteric cult. He also cleared this way a mysterious Greek inscription from the Egyptian city of Sais, recorded by Plutarch, here according to Cudworth's English translation:

> I am all that Hath bee,
> Is, and Shall be,
> and my Peplum or Veil,
> no mortal hath ever yet uncovered.[47]

---

43 Incidentally, in Cudworth's time, the theory that hieroglyphs represented entire words, (later proved to be incorrect), still dominated.

44 Ralph Cudworth: *The True Intellectual System of the Universe*, p. 566.

45 The evolution of Egyptian theology, which continued for some three thousand years, is thouroughly described in Jan Assmann: *Egypt,* Praha: Oikoymenh, 2002.

46 Jan Assmann: *Moses the Egyptian*, p. 125.

47 Ralph Cudworth: *The True Intellectual System of the Universe*, p. 592.

Here speaks a personal divine ego. It says that it is covered with a veil, the veil of the world. Such descriptions of the world, also found in *C.H.*, will turn out to be highly important for Newton, once it becomes clear, what that veil of the world is made of.

Cudworth interprets that veil as the interface between the interior and the exterior; he agrees with Moses that man can see God, so to speak, from behind. He quotes Horapollon: "God is a spirit, removed from the world and streaming through all things from within." And Cudworth therefore concluded: this was the First and Supreme God, "to hen to pan," who contained all things.

We shall see that Newton, in his Scholium Generale adopted this formulation literally.

Cudworth's research casts a new light on god Pan, the old Arcadian god of nature; from his name derives the word *pantheism*. Plutarch told us how sailors on open seas heard a mysterious voice calling "The great god Pan is dead!" Allegedly, it was the crying of ancient demons, fearing that their era was irretrievably gone due to the advent of Christus upon the throne of the Lord of the World.

In other words, at the end of Antiquity, divinity abandoned the world and entered a pure transcendence.

We may now say that Cudworth rescued the so-called *natural theology*, which expanded into deism: here, the ancient interpretation of the divine nature was revived under the assumption that nature is god. (Precise name of that philosophy should be deistic *panentheism*). Ever since the time of Plato and Aristotle, there has been a distinction between god and the world. However, for Egypt, god and nature were still united. God was an unconceivable space that contained all things:

> The true and genuine Idea of God is general, is this, a Perfect Conscious Understanding Being (or Mind) Existing of itself from Eternity, and the Cause of all other things.[48]

After bringing a convincing number of quotations pertaining to the supreme god from a series of ancient authors, Cudworth quotes twenty-three passages from *C.H.* that, he believes, reveal the concept of the highest god in ancient Egypt. He quotes his own hymnic translations of parts of Greek originals of Chapters III, V, VIII, IX, XI, XII, XIII, XV, XVI and a Latin version of Asclepius, too.

48 Ralph Cudworth: *The True Intellectual System of the Universe*, p. 195.

Assmann says that, to this day, Cudworth's translations are overwhelming (*überwältigend*).[49] Their effect was a triumphal return of Hermes Trismegistus and his *cosmotheistic* (universal) religion. This Egyptian religion soon inspired the deists, free masons, modern Hermetics, the universalists as well as the whole Egyptomania of the 18th century, all the way to Mozart's opera *The Magic Flute* and Napoleon's invasion of Egypt.

Let us mention that Cudworth did not attempt to date the documents. He only tried to prove that *C.H.* truly describes the religion of ancient Egypt. He thus reserved the Hermetic tradition for future serious scholarly research.

Newton was likewise convinced that both Egyptian wisdom and monotheism were important in the ancient world. He says:

> Egyptians excelled all other mortals in wisdome. P. 311 Herodot.[50]

This idea must have come to Newton from Cudworth, because had he followed other contemporary sources, he could not have arrived to such conclusions. Likewise, Newton's conviction that the original religious truth, (*prisca sapientia*), revealed by God Himself, accepted by later cultures and transmitted further in a veil of symbols and allegory, was inspired by Cudworth. However, he used another expression, *arcane theology*, which was credited to Origen:

> It hath been already observed out of Origen, that not only the Egyptians, but also the Syrians, Persians, Indians and other barbarian Pagans had, beside their vulgar theology, another more *arcane* and recondite one amongst their priests and learned men.[51]

For the sake of a correct understanding of concepts that Cudworth used and that we shall analyze, we now introduce two special chapters on ancient hermetic philosophy, its contents and its influence.

49 Jan Assmann: *Moses der Ägypter*, p. 128.
50 Out of Cudworth, fN563Z, p. 2 in the original manuscript.
51 Ralph Cudworth: *The True Intellectual System of the Universe* p. 197.

Prisca sapientia

We shall now pay attention to the so-called *prisca sapientia* or *prisca theologia* or *prisca philosophia*:[52] ancient wisdom or theology or philosophy. We shall talk about the origin and spreading of this concept. It will be of utmost importance in our research.

In the year 1463, with the Byzantine scholars coming to the west, the so-called *Texts of Hermes*, in Latin *Pimander* (*Poimandres*), were brought from somewhere in Macedonia. They were collected on the table of Marsilio Ficino, together with original writings of ancient philosophers, such as Plato.[53] Ficino was so impressed by those texts that he preferred to translate *Pimander* earlier than the authentic Plato.[54] To him, it did not matter that the texts were written in bad and evidently recent Greek dialect, the *koine*. He was convinced that they were authentic and that they not only inform us about the old Egyptian religion, but that they were actual initiation texts.

In the introduction to his Latin translation of *Pimander*, Ficino writes:

> Divino itaque opus est lumine, ut soli luce solem ipsum intueamur [...] Mercurius modo sensus, & phantasiae caligines exuit, in aditum mentis se revocans: mox Pimander, id est mens divina, in hunc influit, unde ordinem rerum omnium, & in Deo existentium, & ex Deo manantium, contemplamur.[55]
> (Therefore it is up to the divine light that we should through sunlight see the Sun Himself... Mercury dispels fogs by means of reason and imagination, depending on our mind: soon Pimander, i.e. the divine mind, will fill it, so that we may perceive the universal order that exists in God and from God radiates.)

52 "Philosophers took seriously one particular commonplace of the sixteenth century, namely that the ancients-via revelation dating back to Adam or Noah-had authentic information about theology and science. Knowledge about the latter came to be called the *philosophia perennis*, or perennial philosophy, and is now conventionally called the *prisca sapientia*." Robert Iliffe: *Newton: A Very Short Introduction,* Oxford, New York: Oxford University Press, 2007.

53 Zdeněk Neubauer: Corpus hermeticum Scientiae, in *Logos, sborník pro esoterní chápání života a kultury*, Praha: Trigon, 1997, p. 11.

54 Let us mention that the Greek hermetic writings were known in Europe all the time, but their Latin translation had to wait until the renaissance. According to some scholars, such as Yates or Neubauer, that is what started their extraordinary influence.

55 Marsilio Ficino: *Argumentum Marsilii Ficini Florentini in Librum Mercurii Trismegisti ad Cosmum Medicem, patriae patrem* (manuscript). Zdeněk Neubauer: Corpus hermeticum Scientiae, p. 13.

According to Ficino and his contemporaries that *Corpus Hermeticum* was a true record of the wisdom of Hermes Trismegistos, (in Latin: Mercurius Termaximus), a legendary Egyptian semi-divine philosopher-archpriest.[56] He passed his wisdom, revealed to him by the Supreme God Himself, to posterity.

Via the Egyptian mysteries that original wisdom was supposed to eventually come to Moses and Pythagoras, so that it was a source common to both the Old Testament and to Greek philosophy.[57]

One of those truly fascinated by the ancient Egyptian wisdom was Giordano Bruno. We cannot go deeper into his case: let us just mention that he, more than anybody else at that time, preached a return to Egypt and its magic wisdom, until he was jailed by the Church and, after many years, burned at the stake as an irreparable heretic.

It was not until the humanistic criticism that the victorious advances of the *Corpus Hermeticum* throughout Europe were somewhat mitigated. In 1614, Isaac Casaubon tried to show that the *corpus* is a mixture of many texts, such as the Biblical Genesis, Plato's Timaeus, Psalms, and Gospel after John, early Christian hymns and Gnostic treatises. According to him, there was no ancient Egyptian wisdom in it.

However, many learned men at the time ignored Casaubon. Not only the above mentioned Cudworth, but also the older generation of the Renaissance scholars, such as Robert Fludd[58] or the polyhistor Athanasius Kircher[59] elaborated on the hermetic doctrine. It also survived in the treatises of the Rosicrucians, written certainly by Valentinus Andreae,[60] and it endured in the teachings of Rosicrucian's heirs, the Free Masons. Among them, the hermetic philosophy dominated as the foundation of alchemy, astrology, and kabbalah.

Two other ancient philosophers surfaced in Renaissance Europe as equals of Moses of the Old Testament: Zoroaster and Hermes Trismegistos. Zoroaster was studied by Michael Stausberg[61] in the same way –

---

56 That is the ideal most highly valued also by Isaac Newton: harmonious integration of reason and faith.

57 Neubauer speaks with great respect about the corpus, in spite of calling it a falsum. He credits it with a great contribution to the origin of science.

58 Robert Fludd: *Utriusque Cosmi, Maioris scilicet et Minoris, metaphysica, physica, atque technica Historia*, Oppenheimii 1617–19.

59 Athanasius Kircher: *Œdipus Ægyptiacus*, Amstelodami, 1652.

60 Frances A. Yates: *Rozenkruciánské osvícenství*, Praha: Pragma, 2000.

61 Michael Stausberg: *Faszination Zarathustra. Zoroaster und die Europäische Religionsgeschichte der Frühen Neuzeit*, Berlin und New York: de Gruyter, 1998.

following the *trace of memory*[62] used by Assmann many centuries later. What was *Corpus Hermeticum* for Hermes Trismegistos, were the *Chaldaean Oracula* for Zoroaster.

Western history of memories, *mnemosyné*,[63] encountered Zoroaster through Gemistus Plethon (1355–1454), a Christian who came from the Turkish-occupied Constantinople to Florence. Plethon may be considered the first follower and therefore inventor of the *prisca sapientia* theory. He believed that his own philosophy, Platonism, was a late upshot of the original wisdom, *Urphilosophy*, whose beginnings he placed, obviously following Plutarch, in the period 5000 years before the Trojan War. He believed that the *Chaldaean Oracula* were the oldest remnants of that wisdom.

For Plethon, the *prisca sapientia* became a new religion, of which he felt to be the prophet. That religion aspired to a very noble goal: it was to bring everlasting peace between Christians and pagans. Plethon's effort of rejecting any distinction between the true and the false religion (the core of any monotheistic religion – the Mosaic distinction, i.e. monotheism vs. polytheism) has been the most radical experiment to date.[64] His contemporaries frequently suspected that he was a modern pagan, a *polytheos*. Although all other efforts of reconciliation of Christianity with paganism in some way followed Plethon, they remained on Christian grounds. It is worth mentioning that this radical deflection from Biblical tradition was brought about by following a much older, original tradition: that could have been an advantage, because, according to Plethon, at the very beginning of history there stood the Divine Revelation of all wisdom, which was subsequently gradually lost.

This is why Plethon's new religion was interpreted as a return to the oldest tradition, the one closest to God. This most radical belief was later, in the Florentine Renaissance, only mitigated and headed from heterodoxy toward orthodoxy: thus opposite to the later direction of the 17th and 18th centuries, when orthodoxy (Spencer and Cudworth) headed toward heterodoxy of later deists and Free Masons.

62 Following the trace of memory means to retrace one's steps in history: in the case of Moses, starting with the Cambridge Platonists of the 18th century, Spencer and Cudworth; they got it from the Florentine Neo-Platonists and from Maimonides; he received it from the Greek sources of Manetho; and, at the end, we see the first real monotheist who inspired Moses: i.e., the pharaoh Akhenaten. See Assmann, Jan: *Moses the Egyptian*, Cambridge: Harvard University Press, 1998.

63 See chapter Newton as a Historian where this method, *mnemosyné,* is briefly described.

64 Jan Assmann: *Die Mosaische Unterscheidung*, München und Wien: Carl Hanser Verlag, 2003, p. 108.

Marsilio Ficino (1433–1499) tried to revive Christianity by means of Platonism and Hermes Trismegistos. For him, *prisca sapientia* was not a new religion, but a return to an ancient (=*prisca*) religion, and, at the same time, an original philosophy as a foundation of natural science.[65] He wrote the following, highly influential lines that also contain probably the first available mention of the *prisca theologia*:

> Ille igitur quemadmodum acumine atque doctrina: philosophis omnibus antecesserat: sic sacerdos inde constitutus sanctimonia vita: divinorumque cultu: Universis sacerdotibus praestitit: ac demum adeptus regiam dignitatem: administratione legum rebusque gestis superiorum regum gloriam obscuravit: ut merito ter maximus fuerit nuncpatus. Hic inter philosophos primus: a physicis, ac mathematicis ad divinorum contemplationem se contulit. Primus de maiestate dei: dæmonum ordine: animarum mutationibus sapientissime disputauit. Primus igitur theologiæ appelatus est autor: cum sequutus Orpheus secundas antiqæ theoligiæ partes obtinuit. Orphei sacris initiatus est Aglaophemus. Aglaophemo succesit in theologia Pythagoras quem Philolaus sectatus est diui Platonis nostri præceptor. Itaque una **priscæ theologiæ** undique sibi consona secta ex theologis sex: miro quodam ordine conflata est exordia sumens a Mercurio: a diuo Platone penitus absoluta.[66]

Ficino, too, denies a sharp distinction between monotheism and polytheism, although not along Plethon's neopagan line, but in a broader concept of Christianity: Christianity was in fact anticipated by the older cultures. That approach corresponded to the theory of the *diffusion of the truth*, popular in those times: all ancient religious traditions were believed to derive from a single point, namely from the Biblical Noah. The Bible says:

---

65 Ibid., p. 109.

66 Marsilio Ficino: *Opera omnia*, Basel 1576 (Reprint Turin 1983), p. 1836. He thus surpassed all philosophers in wisdom and learning. As a priest, he laid the foundations for his holy life and surpassed all other priests in worshipping god. Finally, he accepted the royal powers and, by his laws and his deeds, surpassed the glory of the greatest kings. Therefore he was rightly called Trismegistos, three times the greatest. He was the first philosopher who turned away from science and mathematics toward contemplating god. He was the first who lectured on the glory of god, on the orders of demons and on transformations of the soul. That is why he is called the first theologian. He was followed by Orpheus who, in ancient theology, holds second place. Aglaophemus was initiated into the Orphean mysteries. Him in philosophy followed Pythagoras, and, after him, Philolaus, the teacher of our divine Plato. That is why here is a continuous wisdom of ancient theology, rising in splendid order from those six theologians and coming eventually from Mercury and reaching its peak in the divine Plato.

> Noah was a just man and perfect in his generations, and Noah walked with God.[67]

Here we have another example of a single Biblical verse with unexpected consequences, if correctly interpreted. A modern man, unless he is a theologian, does not appreciate the importance of the words "Noah walked with God." Such closeness, even friendship with God, the Old Testament does not grant to anybody else, neither to Moses. This is where the idea of the original revelation of a complete divine wisdom comes from. The Renaissance subsequently worked with that idea under the name of *prisca sapientia*. Its influence was significantly carried over even into the following centuries.

According to that belief, the contents of the divine revelation spread from Noah to his sons and grandsons, and eventually to all lands. Hermes Trismegistos and Zoroaster were then the oldest links in a chain of philosophers or prophets,[68] who were passing that wisdom down from generation to generation throughout the history of this world. That principle offered the opportunity to follow other lines of thought, parallel to the Mosaic one, all the way to the very root, to Noah. For all religions undoubtedly came form God and were then interpreted in different ways by different pious men. That is why the old traditions differ from one another.

Ficino connected the idea of *prisca sapientia* with his belief that ancient *sapientia* was what we would call a form of life: the unity of wisdom, piety and practical abilities, such as healing, care for the soul and philology. The Renaissance idea of a *magus*, who had all those faculties and was able to combine them, believed that Zoroaster and Hermes Trismegistos were the prime examples. Through the union of theology, art, philosophy and natural science, this type of a magus implied the elimination of a Biblical (Mosaic) distinction between the spiritual sphere and the secular one.

Another representative of *prisca sapientia* was Giovanni Pico della Mirandola (1463–1494), called Priceps Concordiae.[69] It is interesting that he had made a complete turn-around during his short life. At first an enthusiastic preacher of *prisca sapientia*, he suddenly rejected it and started calling it *prisca superstitio*. The increasing orthodoxy of the 17th century soon followed him in this opinion.

67 Genesis 6,9.

68 Newton believed that he was a link in that continuous chain.

69 Jan Assmann: *Die Mosaische Unterscheidung*, p. 110. The founder of harmony.

Agostino Steuco (1497–1548) introduced his own *philosophia perennis*,[70] based on Ficino's concept of ancient theology. He, too, tried to abolish the distinction between a correct and incorrect religion and to replace it with a mutual harmony of wisdoms of all nations and of Christianity. He also engendered the idea of an original divine revelation which was then further communicated by the greatest of the wise men. This process of diffusion, which introduced a certain ramification into various cultures from a single spot of a primeval religion, was understood as progress by scholars of the Ficino generation, but shows some marks of decay as early as Steuco. The concept of *prisca sapientia* brought the realization that the process of transfer of the original message from generation to generation corrupts the message. Its radical rejuvenation then demands a return to the original sources.

Steuco believed that Christianity was the only creed that had not succumbed to the decadent forms of other religions. This deprivation variant of the *diffusion theory* finally, in the 17th century, became the dominant cultural model.

Plethon perceived his vision of a new religion under the influence of the conflict between the Eastern and Western Church and between Christianity and Islam, and Steuco analogously felt the tension between Catholicism and Protestantism as irreparable. So far, all those Renaissance returns to the past, whether to Zoroaster, Hermes Trismegistus, or Plato, to astrology or to kabbalah, were in the services of attempts toward a peaceful toleration of other cultures. The common source of all religions was to enable man to overcome contemporary struggles, because they offered a return to pre-Mosaic prophets of divine wisdom.

Of considerable consequence in our research is Francisco Patrici da Cherso (1529–1597),[71] latinized to Franciscus Patricius and in Croatian referred to as Frane Petrić, who was a Croat from the island of Cres. In his *Nova de Universis Philosophia* he also tried to unite a rejuvenated *prisca sapientia* with non-confessional Christianity. He attempted to combine four existing philosophies with a fifth one, his own.

As a true Renaissance man, he was involved in many subjects: besides philosophy, he worked in mathematics, poetry, music, botanics, physics, even the art of war. He worked out his own deprivation theory for linguistics, introducing the idea of an original language, *prima lingua*[72]:

70 Ibid., pp. 110–11. Eternal philosophy.

71 Ibid., p. 111.

72 According to Aleida Assmann, that original language, lost at the Tower of Babylon, operated on the principle of *immediate meaning*: direct and natural use of a sign code which, by

according to him, the original, immediate meaning of words was still alive only in foreign magic formulas, in *onomata barbara*, and gave them their power. Egypt, with its as yet undeciphered hieroglyphs, was of course central in this search for the original language. Since Aristotle's theory of language and letters as a means of communicated message was still observed by educated people, we may comprehend Patrizi's idea of immediate meaning as a Platonic, counter-Aristotelian opposition. Patrizi was respected as an acute and original observer of nature. He inaugurated the theory of light, movement of the Earth and reproduction of plants. In the spirit of neo-platonic emanatism, he believed the celestial *prime light* was the beginning of all being. This idea may have influenced Newton (and, of course, Paracelsus).

It is very important for us that Patrizi was one of the first to publish the hermetic writings in print. The very first edition, entitled *Poimandres*, was published as early as 1471. Incidentally, it was the hermetic opus, for long time forgotten and only recently translated from Greek to Latin, which immediately benefitted from Gutenberg's invention of the printing press. Such books could spread faster and were more affordable.

We know of several editions of the hermetic opus. Of those published by Patrizi, we have the Ferrara edition of 1591 and the Hamburg edition of 1593. The latter is of consequence for our work, because Newton had a copy in his own library.[73]

Corpus Hermeticum I.

Greek writings, usually called *Corpus Hermeticum* are a mysterious and highly influential document of mid-platonic philosophy. Today we believe it originated in second-century Alexandria. It is not quite clear how *C.H.* came to Europe, except for the last stage of its transition, i.e. transported by the exiled Byzantine monks. The earliest mention of *C.H.* is by Origen, Augustin and Stobaeus speak of it in more detail, it is also mentioned by Climent of Alexandria and by Tertulian, and

convention, defines the relation between the item and its symbol. See Aleida Assmann: *Die Legitimität der Fiction*, München, 1980.

73 According to: John Harrison: *The library of Isaac Newton*, p. 212, Newton had Patricius' book in his library.
1267 Franciscus Patricius: *Magia Philosophica, hoc est F. Patricii summi philosophi Zoroaster & eius 320 Oracula Chaldaica. Asclepp Dialogus. &Philosophia magna Hermetis Trismegisti ... Latine reddita.* Hamburgi, 1593.

frequently quoted by the Neo-Platonists. Nowadays, the basic edition is that of Festugiere.[74] It contains the fourteen old Greek books, and several others, among them mainly the long *Asclepios* present only in Latin; further the "Emerald Tablet," known only from a seventh-century Arabic book.

Neither final analysis, nor explanation of all meaning is yet completed. Harmonizing the conflicts among different books is not easy. A recent surprise was the discovery of some of the *C.H.* books among the Gnostic documents found at Nag Hammadi in Egypt, e.g., the chapter *On Number Eight and Seven*, similar to *C.H.* Treatise XIII; and the *Prayer of Thanks to Asclepius*. Many scholars admit that *C.H.* may derive from a much older canon and that it may be a mid-platonic expression of a much older wisdom. Kratochvíl, e.g., writes:

> The concept of the relation of the world, symmetry and overall mutual connection also seem to be archaic. [...] We do not want to look for old-Egyptian influence. But we know that fundamental paradigms of learning and thinking may survive incredibly long in a given cultural and geographic region, even without any obvious continuity and intentional transmission; almost as a kind of *genius loci*. [...]
> We do not assume that the older layer of the hermetic canon came from ancient times and was translated into Greek in its present form. However, their intellectual structure and their function may point toward a sophisticated tradition of wisdom that had to wait for the contact with the neo-platonic philosophy, in order to get recorded. One of the purposes of those texts may be a preparation of the mind for thinking: they may prepare consciousness for something serious to think about.[75]

The so-called *hermetic writings* are addresses of Hermes Trismegistus to his son Tat, or to Asclepius (son of god Apollo) or to king Amun. Hermes was a Greek god, and is taken for a parallel to the Egyptian god Thoth. Hermes Trismegistos, from the contemporary perspective, is a mythical demigod. But whether we see him as a Hellenized Egyptian god, or an ancient arch-priest, we have to credit him with considerable power; and that power was certainly much greater in the Renaissance and at the dawn of modern times. By not being precise in the using of terms, *C.H.* is a threshold between mythical and philosophical thinking,

74 André-Jean Festugière: *Corpus hermeticum* I–IV, Paris: Belles Lettres, 1991.

75 Zdeněk Kratochvíl: *Prolínání světů,* Praha: Herrmann a synové, 1991, p. 68. Translated by the author.

sometimes called *sapiential literature.* It gives them a kind of power that still, two thousand years later, tells us about the wisdom of god Hermes.

Hermes, in Latin Mercurius, Egyptian Thoth-Jehuti[76] was a god full of opposites, both revered and scorned. He was a leader of souls, a messenger of gods, a personification and an invisible controller of interfaces of all kinds.

He could overcome any horizon[77] and also could lead all others over any horizon. It could be a physical barrier, for instance a battle line or the difference between wealth and poverty[78] or the difference between honesty and fraud. However, it may also be the difference between the visible and the hidden, known and incomprehensible, life and death. It was also this god's privilege to cross and transfer, (also to cheat), and to interpret: to this day, we use the term *hermeneutics.*

Thus the general purpose of all hermetic texts is to carry the reader over the horizons, over seemingly unsurpassable boundaries: the boundaries of sensual perceptions, everyday experiences on the one hand, and, on the other hand, true wisdom.[79] Hermetic texts do not excel by depth of thought, or by accuracy of expression. Compared with Plato, they are decadent. Nevertheless, they possess a value that constantly demands new translations and new exegesis; new efforts to overcome their darkness, internal contradictions and inaccessibility.[80]

According to Radek Chlup,[81] the secret of all those documents is in their special relation between comprehension and being itself; when Chlup wants to express the uniqueness of the Hermetic writings, he refers to Plato and his explanation that philosophy acts on five levels. As far as the fifth level is the ultimate goal, and all else is meaningless without it.

76 At some time, antique gods became translatable from one culture into another. It happened when gods were no longer tied to particular human clans and territories but were identified with their several linguistic, iconic and cultic responsibilities. Their identity became semantic. This is evident from ancient agreements between states, or from the Golden Ass by Apuleius of Madaura, where we find a list of names that various nations give to a goddess corresponding to the Egyptian Isis: Pessinuntia, Athéna, Proserpina, Deméter, Héra, etc. See Jan Assmann: *Moses the Egyptian*, pp. 44–48.

77 The planet Mercury is a precise personification of his function: always close to the horizon, as if he ruled it.

78 Hermés was also god of merchants, and even of thieves.

79 For the sake of completeness, we mention that passing over the border can mean either looking there, so that we move our horizon, or we ourselves cross over. And that is the true meaning of Mercury: not only seeing across the border, but also stepping over it.

80 Using Jung's terminology, one of the principal Hermes' horizons is between consciousness and unconsciousness. Perhaps, with a little exaggeration, we may say that therein lies the boundary between rational thinking and value thinking.

81 Radek Chlup: *Corpus hermeticum*, Magisterial thesis, Philosophical Faculty of Charles University in Prague (undated).

Briefly: the first three levels are name (*onoma*), definition (*logos*) and image (*eidolon*). Plato calls the fourth level knowing, thinking and having the correct opinion about things, i.e. internal understanding, which takes place in a man's soul. But then there is the fifth, highest level, for which Plato does not have a right word and, therefore, has difficulties trying to explain that he means a unification of understanding and being, subject and object, perception and opinion, because it was not yet only observed, but put and lived together.[82]

Therefore, the fifth level could be described as the situation when man, by his comprehension, is internally and deeply changed, transformed. According to Plato, to fulfil the purpose of philosophy, the perfect rational work is not enough; it is moreover necessary to be in contact with deeper, extra-rational roots of the soul, with the experiences of man as a whole, including his emotionality. To overwhelm the listener or reader by a multitude of correct information or brilliance of rational thought is not enough for a transition onto the fifth level.

According to Chlup, the ancient hermetics made a risky step by skipping the fourth level; the level of philosophy: precise logics, dialectics, systematic thinking, definitions etc. They only concentrated on the fifth level, for which they even had a new name: *gnosis*.

Nowadays, we use the term *gnosis* for a multitude of sects; nevertheless, *gnosis* has one fundamental characteristic that distinguishes it from other systems, such as Christianity: it is the concept of the way toward salvation. In Christianity, salvation is given by God's Grace through Jesus Christ, and accessible via the Church, which acts as Christ's mystical body in this world. Whereas in *gnosis*, salvation is achieved only through individual recognition. However, it is not just any recognition: it must be the recognition of man's unity with god. That is precisely Plato's recognition that is needed for a transition into the fifth level. Expressed in the language of Hermes, it is a dramatic change of horizon. It exposes a new, different world to man.

Another characteristic of *gnosis* is its individuality: for salvation, *gnosis* does not need any organization, such as the Church. We have to keep in mind this Gnostic approach to salvation as long as we want to study Newton's opus: for Newton, Jesus was just a man. And Newton did not depend on Church at all.

Let us mention here the opinion of a great authority on *gnosis*, C. G. Jung:

82 Ibid., p. 59.

Unlike the majority of scholars of his day, Jung never interpreted Gnosticism as Christian heresy of the 2nd and 3rd century. Neither did he pay any attention to the endless debates about a possible Indian, Iranian, Greek or other origin of Gnosticism.

Earlier than any authority on Gnostic studies, Jung understood who the Gnostics really were: visionaries who created original, primary studies from mysteria which he calls *the collective unconscious*. When, in 1940, he was asked whether Gnosticism was a philosophy or a mythology, Jung answered in all earnestness that the Gnostics were interested in real, original images and that they were not any kind of syncretic philosophers, as many think.

He noticed that Gnostic images originate even now in man's internal experiences in connection with *individuation* of human *psyché*; that, for him, was a corroboration of his opinion that the Gnostics created true, archetypal images that are known to last and exist regardless of time or historical events. In Gnosticism, Jung discovered a powerful, primary and original expression of human mind, heading toward the most profound end of the soul: the reaching of the wholeness.[83]

To this day, hermetic writings still possess a power to transform those who seriously study them. They truly provide answers we seek of them. Treatises of the *C.H.* help us recall old experiences, mysterious intuitive cognition and *silent knowledge*. From that viewpoint, according to Neubauer, "*C.H.* is a truly original wisdom, necessary supposition of any comprehension and understanding of the world, of the facts of cosmos."[84] The fundamental hermetic motive, the relation between the hidden and the revealed, is always philosophically crucial. There is something that cannot be directly pointed out, that can be comprehended in context only: namely, it is the cosmos as a whole, with all its potential connectivity. As Karl Jaspers claimed:

All our knowledge always remains within the world, never comprehends the world as such. World is not an object of research, but, as Kant says, it is an idea that leads research and gives it a unity. [...] The price for great universal concepts that impress us so much, particularly in physics and cosmology, is always an abstraction of an enormous number of otherwise meaningful facts which, here, no longer deserve mention.[85]

---

83 Stephan A. Hoeller: *C. G. Jung a gnóze*, Praha: Eminent, 2006, p. 55.

84 Zdeněk Neubauer: *Corpus hermeticum scientiae*, p. 15.

85 Karl Jaspers: *Chiffren der Ttranscendenz* (in Czech: *Šifry transcendence*, Praha: Vyšehrad, 2000, p. 39).

Since there are several possible transitions across the horizon, and since the connection between the revealed and the hidden is multiplex, Hermes often uses his controversial signs: he may be profoundly wise, but also irreverent, even fraudulent. It is understandable: the horizons are blurry, movable and infinitely many. That is why it is so difficult in hermetic thinking to distinguish between a profound idea and a fraud. It is somewhat similar to a game of hide-and-seek.

For better understanding, we include a belletristic analysis of god Hermes from the Nobel laureate Thomas Mann's tetralogy, *Joseph and his Brothers*:

> The mythical popularity that Joseph acquired and that it had probably always been his nature to acquire was based above all on the shimmering mixed character and ambiguity – mirrored by the laughter in his eyes – of his measures, which functioned, as it were, in two directions at once, combining in a thoroughly personal way his various purposes and goals with a kind of magical wit. We speak of wit because this principle has its place in the little cosmos of our story and early on the statement was made that wit is by nature a messenger who goes back and forth, a nimble ambassador between two opposing spheres and influences – for example, between the forces of the sun and moon, the father's legacy and the mother's legacy, between the blessing of the day and the blessing of the night – indeed, to put it in direct and all-inclusive terms, between life and death. Such mediation, so slender and agile as it merrily goes about reconciliation, had never found real expression in any divinity in the land where Joseph was a guest, in the Land of Black Earth. Thoth, the scribe and guide for the dead, the inventor of so many clever things, came closest to such a figure. Only Pharaoh, before whom all divine matters were brought from far and wide, had knowledge of a more perfected version of this divine character, and the grace that Joseph had found before him was due primarily to the fact that Pharaoh recognized in him the traits of that rascal child of the cave, that lord of tricks, and had quite rightly told himself that a king could wish for nothing better than to have as his minister such a manifestation and incarnation of this profitable divine idea. The children of Egypt became acquainted with this winged figure through Joseph, and if they did not include him in their pantheon it was only because that place was already taken by Djehuti, the white monkey [...][86]

86 Thomas Mann: *Joseph and His Brothers*, translated by John E. Woods, Westminster, MD, USA: Knopf Publishing Group, 2005, p. 1439. Also: http://site.ebrary.com/lib/natl/Doc?id=10078782&ppg=1479.

For the sake of better understanding and as an introduction into this problem, we provide here the complete text of the fifth book of the *Corpus Hermeticum* in John Everard's English translation:

The Divine Pymander in XVII books. London 1650. This was translated by John Everard from the Ficino Latin translation.[87]

The Fifth Book.
"That God is not Manifest and yet most Manifest."

1. This Discourse I will also make to thee, O Tat, that thou mayest not be ignorant of the more excellent Name of God.
2. But do thou contemplate in thy Mind, how that which to many seems hidden and unmanifest, may be most manifest unto thee.
3. For it were not all, if it were apparent, for whatsoever is apparent, is generated or made; for it was made manifest, but that which is not manifest is ever.
4. For it needeth not to be manifested, for it is always.
5. And he maketh all other things manifest, being unmanifest as being always, and making other things manifest, he is not made manifest.
6. Himself is not made, yet in fantasy he fantasieth all things, or in appearance he maketh them appear, for appearance is only of those things that are generated or made, for appearance is nothing but generation.
7. But he is that One, that is not made nor generated, is also unapparent and unmanifest.
8. But making all things appear, he appeareth in all and by all; but especially he is manifested to or in those things wherein himself listeth.
9. Thou therefore, O Tat, my Son, pray first to the Lord and Father, and to the Alone and to the One from whom is one to be merciful to thee, that thou mayest knowest and understand so great a God; and that he would shine one of his beams upon thee In thy understanding.
10. For only the Understanding sees that which is not manifest or apparent, as being itself not manifest or apparent; and if thou canst, O Tat, it will appear to the eyes of thy Mind.
11. For the Lord, void of envy, appeareth through the whole world. Thou mayest see the intelligence, and take it in thy hands, and contemplate the Image of God.

87 John Everard: *The Divine Pymander*. See http://www.levity.com/corpherm.html.

12. But if that which is in thee, be not known or apparent unto thee, how shall he in thee be seen, and appear unto thee by the eyes?
13. But if thou wilt see him, consider and understand the Sun, consider the course of the Moon, consider the order of the Stars.
14. Who is he that keepeth order? for all order is circumscribed or terminated in number and place.
15. The Sun is the greatest of the Gods in heaven, to whom all the heavenly Gods give place, as to a King and potentate; and yet he being such a one, greater than the Earth or the Sea, is content to suffer infinite lesser stars to walk and move above himself; whom doth he fear the while, O Son?
16. Every one of these Stars that are in Heaven, do not make the like, or an equal course; who is it that hath prescribed unto every one, the manner and the greatness of their course!
17. This Bear that turns round about its own self; and carries round the whole World with her, who possessed and made such an Instrument.
18. Who hath set the Bounds to the Sea? who hath established the Earth? for there is some body, O Tat, that is the Maker and Lord of these things.
19. For it is impossible, O Son, that either place, or number, or measure, should be observed without a Maker.
20. For no order can be made by disorder or disproportion.
21. I would it were possible for thee, O my Son, to have wings, and to fly into the Air, and being taken up in the midst, between Heaven and Earth, to see the stability of the Earth, the fluidness of the Sea, the courses of the Rivers, the largeness of the Air, the sharpness or swiftness of the Fire, the motion of the Stars; and the speediness of the Heaven, by which it goeth round about all these.
22. O Son, what a happy sight it were, at one instant, to see all these, that which is unmovable moved, and that which is hidden appear and be manifest.
23. And if thou wilt see and behold this Workman, even by mortal things that are upon Earth, and in the deep. Consider, O Son, how Man is made and framed in the Womb; and examine diligently the skill and cunning of the Workman, and learn who it was that wrought and fashioned the beautiful and Divine shape of Man; who circumscribed and marked out his eyes? who bored his nostrils and ears? who opened his mouth? who stretched out and tied together his sinews! who channelled the veins? who hardened and made strong the bones! who clothed the flesh with skin? who divided the fingers and the

joints! who flatted and made broad the soles of the feet! who digged the pores! who stretched out the spleen, who made the heart like a Pyramis? who made the Liver broad! who made the Lights spungy, and full of holes! who made the belly large and capacious? who set to outward view the more honourable parts and hid the filthy ones.

24. See how many Arts in one Matter, and how many Works in one Superscription, and all exceedingly beautiful, and all done in measure, and yet all differing.
25. Who hath made all these things! what Mother! what Father! save only God that is not manifest! that made all things by his own Will.
26. And no man says that a statue or an image is made without a Carver or a Painter, and was this Workmanship made without a Workman? O great Blindness, O great Impiety, O great Ignorance.
27. Never, O Son Tat, canst thou deprive the Workmanship of the Workman, rather it is the best Name of all the Names of God, to call him the Father of all, for so he is alone; and this is his Work to be the Father.
28. And if thou wilt force me to say anything more boldly, it is his Essence to be pregnant, or great with all things, and to make them.
29. And as without a Maker, it is impossible that anything should be made, so it is that he should not always be, and always be making all things in Heaven, in the Air, in the Earth, in the Deep, in the whole World, and in every part of the whole that is, or that is not.
30. For there is nothing in the whole World, that is not himself both the things that are and the things that are not.
31. For the things that are, he hath made manifest; and the things that are not, he hath hid in himself.
32. This is God that is better than any name; this is he that is secret; this is he that is most manifest; this is he that is to be seen by the Mind ; this is he that is visible to the eye; this is he that hath no body; and this is he that hath many bodies, rather there is nothing of any body, which is not He.
33. For he alone is all things.
34. And for this cause He hath all Names, because He is the One Father; and therefore He hath no Name, because He is the Father of all.
35. Who therefore can bless thee, or give thanks for thee, or to thee.
36. Which way shall I look, when I praise thee? upward? downward? outward? inward?
37. For about thee there is no manner, nor place, nor anything else of all things that are.

38. But all things are in thee; all things from thee, thou givest all things, and takest nothing; for thou hast all things and there is nothing that thou hast not.
39. When shall I praise thee, O Father; for it is neither possible to comprehend thy hour, nor thy time?
40. For what shall I praise thee? for what thou hast made, or for what thou hast not made! fur those things thou hast manifested, or for those things thou hast hidden?
41. Wherefore shall I praise thee as being of myself, or having anything of mine own, or rather being another's?
42. For thou art what I am, thou art what I do, thou art what I say.
43. Thou Art All Things, and there is Nothing Else Thou art not.
44. Thou Art Thou, All that is Made, and all that is not Made.
45. The Mind that Understandeth.
46. The Father that Maketh and Frameth.
47. The Good that Worketh.
48. The Good that doth All Things.
49. Of the Matter, the most subtle and slender part is Air,
of the Air the Soul,
of the Soul the Mind,
of the Mind God.

III. Newton the Theologian and Historian

He answered and said unto them: Well hath Esaias prophesied of you hypocrites, as it is written. This people honoureth me with their lips, but their heart is far from me. Howbeit in vain do they worship me, teaching for doctrines the commandments of men.
Mark 7,6–7[88]

A number of prominent scholars[89] have already analyzed Newton's theological opus. As long as we are looking for extra-scientific inspirations in Newton's scientific work, a significant part of Newton's theological and historical opus is outside our sphere of interest – those works were produced in the last years of Newton's life, when he no longer worked on problems of natural science. However, Newton's theological quest spanned 60 years and many theological problems were on his mind for all of his life, it is necessary to always keep in mind the omnipresence of God in all of Newton's thought.

God was Newton's measure of value. God was his motivation, the alpha and omega of his entire endeavour. Newton searched for God's Truth. God as an absolute value was not only the framework of his research: Newton also conducted his theological thinking to the limit, i.e. all the way to a mathematical formulation:

> And God said: Let there be lights in the firmament of the heavens, to divide the day from the night; and let them be for signs, and for seasons and for days, and years. And let them be for lights in the firmament of heavens, to give light upon the earth.[90]

88 King James Bible (all quotations).
89 Frank E. Manuel, Robert Iliffe, Steven Snobelen, etc.
90 Genesis 1,14.

This single short sentence from Genesis is the basis for Newton's entire scientific endeavour: the problem of space and time, and the mathematical formulation of celestial mechanics controlled by gravitation, as well as research on light.

Newton, like many of his contemporaries, searched for truth in two domains of God's work: in the sphere of Creation (of the world), thus in natural philosophy, and further in the sphere of God's Word (Bible). God's Work and God's Word are the complete plan of Newton's scholarship. It focused on knowing and worshiping God the Creator, and served as Newton's defence against the growing atheism. While we try to find the sources of modern science, we have to pay attention to those disciplines, such as theology and alchemy that do not contribute to the material side of science, but operate as their framework. For a pious man such as Newton to study this world, created by God, was not just rational labour: it was a kind of a prayer.

While Newton was a deeply pious man, his theological writings reveal his non-conventional, even heterodox opinions. That, by itself, was not unique: many points in traditional dogmatism, both Catholic and Protestant, were always in conflict with the reasoning of rational Christians. And it was unnecessary to have a deep theological education: to have the Bible and to be able to read was all that was needed. And, of course, Newton was one of those precisely thinking Christians. They all believed that the original clear message of Jesus had been later corrupted beyond recognition and that it was necessary to rectify matters by returning to the Bible. We shall try to show that some fundamental points in Newton's heterodoxy may have been inspired by the precise reading of the Bible itself.

In this day and age, there may be some difficulty with the subject of this chapter, so it may help if we clarify matters. Christianity follows from the teachings of Jesus Christ; however, in the strict sense of the word, he was not the founder of the Christian religion. His message was elaborated on, recorded and spread throughout the world by the apostle Paul, by the Gospel writers, and later followers. The Christian belief is based on those sources, is formulated by the teachings of the Church, the so-called *dogma*, and is studied, expanded, and interpreted through a science called theology.

We shall proceed in like manner: first, we shall discuss faith and theology. Afterwards, we shall study at least the main points of Newton's heterodoxy and confront them with modern Protestant theology. At the end of the chapter, we shall mention Newton's historical work.

About Theology

The word "theology" first appears in Plato's *Republic*, where it has the meaning, a bit odd for us, of instructions on how to talk about the gods. As a science, it is first mentioned by Aristotle.

However, Christian theology is not – as some might conclude from the name itself – a science about God. Already the ancient philosopher Poseidonius (c. 135–50 BC) defines theology as follows: *epistémé tón theión kai anthrópión*,[91] knowledge of the divine and human (things).

Poseidonius is interested not only in divine matters, but also in human ones. This is a serious step away from the concept of Aristotle's metaphysics. Aristotle's concept is still retained by Thomas Aquinas' definition: *theologia est scientia de Deo rebusque divinis*, theology is a science about God and things divine, where *divinum* is used for an objective God, while human things are absent. Poseidonius did not yet treat God as an object. This step, treating God as an *object*, became the cardinal error of Christian theology: God is not, and cannot be an object, cannot be cast ahead of the subject as a helpless thing to be analyzed. God is the subject, an absolute Subject.

Today, the object of Christian theology is God in His revelation,[92] where revelation means mainly the revelation of Jesus. Theology is one of the possible bridges between mythical and rational thinking. It attempts to provide a rational treatment of values, i.e. of matters of a higher level of emotion. Values cannot be comprehended within the strictly objective thinking limited to the antithesis of *subject vs. object*. By carrying a value, the object implies emotions, restructures our mind, plays its role in the mental stability of the subject, and thus becomes an internal part of it. An impartial rational thinking may thus be significantly transformed.[93]

A general belief in God is not characteristic for a Christian: against the background of the Old Testament belief in God, he mainly confesses Jesus Christ as his Saviour. Traditional Christianity even accepts Jesus as the so-called second Divine Person; and, together with the third Person, the Holy Ghost, it worships the *Holy Trinity*.

The concept of faith can be understood in two ways: as the content, i.e. that which is believed, *fides quae creditur* (faith which is believed =

---

91 Zdeněk Kratochvíl: *Prolínání světů*, Praha: Herrmann a synové, 1991, p. 15.

92 David Tonzar: *Teologická propedeutika*, Praha: Karolinum 1999, p. 11; Ivana Noble: *Po Božích stopách*, Praha: Centrum pro studium demokracie a kultury 2004, p. 14.

93 Irena Štěpánová: "Isaac Newton, oheň a led," in: *Sborník přednášek Akademie biskupa Mikuláše 2006/07*, Praha, Blahoslav, 2007.

sentences) and may be confessed purely formally without any practical results; or as an act, something that works and is manifested by its results, *fides qua creditur* (faith through which, or by means of which, it is believed = transformative action. Let us remember Plato's fifth level of transformation of the human soul: that is exactly what Plato has in mind.)

In its original Biblical meaning, faith is not limited with verbal formulations and speculations. Biblical faith is an experience with a transformational power. It is based much deeper than thinking.

In this connection, let us remember the semantic genius of the old Hebrew language: the word EMN means both *truth* and *faith*. It means that, by his life, man demonstrates what he really believes, what is his internal truth. Much later, the same would be stated by Martin Heidegger: our own existence is a subconscious and practical interpretation.[94]

In the Biblical sense of the word, man does not decide his faith: man is overpowered by his faith. That means that the source of religious experiences is in the emotions of a higher class; theology as a doctrine is secondary, reflective.

Since its origin, theology has had a particular relation to philosophy. Both are in mutual tension, problematizing each other. In many respects, theology is similar to philosophy: it has a similar structure. However, it does not *build on virgin ground*. Theology has particular axioms that follow from its fundamental concept. It is tied to the text of the Bible, to a particular Church and its denomination, and is the interpretation of a particular fundamental experience: experience of faith in Christ and God. They are carriers of an absolute value.

From the very beginning, Christian thinkers had to, and wanted to, stand up to Greek philosophy: the Church's survival was at stake. They had to confront not only the contemporary Hellenistic way of thinking, but, little by little, they ventured to measure themselves against those half-forgotten summits of Greek philosophy, such as Plato (Augustine) and Aristotle (Thomas Aquinas).

And so it happened that the Greek way of thinking controlled the Christian religion from within for many centuries. On the one hand, the magnificent development of Greek speculative ways prepared a new European intellectual age, which includes modern science. On the other hand, however, the very goal of theology – accepting Christ and His message through metaphysical thinking – failed[95]: that message was irrep-

94 Martin Heidegger: *Sein und Zeit*, p. 170 in the Czech edition: *Bytí a čas*, translated by Ivan Chvatík et. al., Praha: Oikoymenh, 2002.

95 Ladislav Hejdánek: *Filosofie a víra*, Praha: Oikoymenh, 1999, p. 119 et seq.

arably damaged. In theology, theorizing, which lacked the necessary real experience with God and Christ, got the upper hand. Thus it happened that for centuries, theology forgot about the category of experience.

Orthodoxy did not trust experience. Individual experiences were suspect. There was even suspicion whether man understands his own experiences. (We may find this opinion in Karl Barth's writings in the 20th century.)

Not until the middle of the 20th century did theology receive intellectual tools that helped it comprehend the Biblical message: it happened under the influence of existentialism and personalism (e.g. Karl Jaspers and Martin Buber), where the barrier between the subject and the object is overcome.

Ridding theology of metaphysics, removing fruitless verbal speculations that lack living faith has been Christianity's perennial problem. Excellent contributions to this discussion are books by Claude Tresmontant.[96] A religious paradigm that does not address anyone does not have the necessary sentimental element revealing authentic emotions, the very source of spiritual energies, and must sooner or later fail: it does not internally communicate.

The Judeo-Christian concept of God, based on Old Testament theology, assumes that God is separate from the world, from His creation. God has an external relation with the world. And, further, God has His own existence, independent of the world. Only this separate and independent existence permits the possibility of the Last Judgment, life after death, the Kingdom of Heaven, etc. That is the true Biblical monotheism, a strict distinction between God and the world, the so-called divine "transcendence." Although passages such as "God in this world," or "God in us" can be found even in the Orthodox tradition, and, perhaps, the entire Christian mysticism stands on such concepts, those passages always hold a poetic license; they are not meant to be taken literally, otherwise Biblical monotheism would simply collapse.[97]

Contrary to this Jewish and Christian *transcendent* God is an *immanent* God, such as the concept of God preached by Giordano Bruno. In his religious system, *pantheism*, everything, the entire universe, is a physical revelation of God Himself. God and His creation are identical. God does not have an independent existence; therefore He is not a person. It is hard to pray to such a God. Life after death is out of the question, etc.

---

96 Claude Tresmontant: *Bible a antická tradice*, Praha: Vyšehrad, 1998.

97 Ivo Tretera: *Nástin dějin evropského myšlení*, Litomyšl: Paseka, 2002, p. 253.

We should yet remark something about the stance of the early Church toward Jesus Christ (Christology): it underwent a long and complex process of changes, with its final formulation in the year A.D. 325. In that year, at the Council of Nicea presided by Emperor Constantine, the Church turned away from worshiping a single God and adopted the worship of the *Trinity* (the Father, the Son and the Holy Ghost); Jesus Christ was voted God (!).

The conflict between Arius and Athanasius, the protagonists of the two opposing parties, exists in western Christianity to this day. Simply stated, the Arians cared about the faithful Christians, for whom Jesus as a Divine Messenger brought faith. Their opponents, the Athanasians, cared first about the Church to which Christ entrusted the power of salvation. In effect, it means that there is no salvation outside the Church.

But that puts the argument of Jesus' divinity into another realm: the question is whether it is possible to follow Jesus Christ individually, or whether a man is confined to merely being a sheep, obedient to the hierarchical Church.

The Church's doctrine on the Trinity forever conflicted with the reasoning Christians. No wonder that modern Christians, under the influence of humanism, the Renaissance and the critical tendencies of the Reformation, tried to dissociate themselves from the belief in the Trinity. The Reformation in fact retained the ancient Christian dogma of the Trinity. The concepts of anti-trinitarianism and unitarianism are self-explanatory: they tend to replace the belief in three metaphysical divine persons with the belief in a God of one person. In Newton's time, a frequently used term for this position was "socinianism," derived from the name of the principal propagators of Unitarianism, the Italian Faustus Sozzini (1539–1594) and his uncle, Laetius Sozzini (1525–1562).

Some individuals, regardless of other differences, attacked the Trinity at its very root: they denied Jesus Christ as God. For some of them, ridding religion of both the divinity of Jesus and of the Trinity was the most important step toward the purification of Christianity. Those people stressed the belief in a single personal God, and Christ was recognized as a mere human, albeit a prominent one. [98]

98 The chapter about anti-trinitarianism has been prepared according to the scripts of *Dějiny dogmatu*, Praha: Ústřední rada CČS.

Newton as a theologian of his times

Newton belonged to the Church of England.[99] Both his stepfather and his uncle were Anglican priests. His theological opus is typical for its period. It is quite voluminous: extant writings contain some one million words. In the past, many scholars, mainly the positivistically oriented ones, could not forgive his "extra-scientific" interests. They apologetically employed pseudo-explanations, such as Newton's senility, graphomania, love of his own beautiful handwriting etc. At present, those extra-scientific interests are the subject of solid scholarship.

As a child, Isaac Newton received a strict and solid Puritan education. All his life, he suffered from anxiety, an inferiority complex, and a deep feeling of guilt. At the same time, he was convinced of his excellence, even of a calling to finally uncover all the secrets of both the world and the Bible. (It may be understood as a psychological compensation). Newton knew the Bible thoroughly: he studied it in Latin, Greek, English, and, with a dictionary, Hebrew.

Let us look at some interesting points in his theological studies. We shall start with Newton's ideas about Jesus Christ, whence the beginnings of his heterodoxy originated. Afterwards, we shall look at his criticism of the Church. And, finally, we shall study his position toward Moses, who had a defining influence upon Newton's attitude toward science.

Newton's idea about Jesus

It underwent a long development. During his active alchemistic period, he was interested in Christ's role in the creation, and around 1673 he found himself very close to the Arians: Christ was understood as an angelic being of universal importance, mediating and exacting God's will in this world, and, finally, accepting a human body as the most perfect

99 The Church of England was established by King Henry VIII after a conflict with the Pope. In 1531, with the help of the Parliament, he dissociated the Anglican Church from Rome and became its head. The Archbishop of Canterbury found support among the theologians of the German Reformation, which then gained a strong spiritual influence in England. One of the direct consequences was an emphasis on the Bible (unlike the Catholic tradition, where the Church tradition has the same value as the Bible). A thorough knowledge of the Bible has had a long tradition in the English nation.
The Anglican Church of those times stressed a living, simple Christianity, a pious heart, modesty and moral dignity. It put emphasis on promoting Christianity in the life of the nation, among other things, by establishing clubs. In 1698, the "Society for Promoting Christian Knowledge" was established. That society always professed a strong social commitment and accomplished much in the field.

human. Today, we would likely call his approach the *cosmic concept of Christ.*[100]

However, Newton later abandoned those Arian ideas about Christ and talked about him as a man. He always called him the "man Jesus". He thus became an anti-trinitarian, and therefore a heretic. Of course, nowadays we comprehend that attitude, since it is a recurring one: e.g., among the Czech Catholic liberals of the early 20th century, it resulted in the foundation of a new Church. Likewise, T. G. Masaryk, himself a Unitarian, used to claim: "For me, Jesus is everything."

Recently, a series of theologians have tried to demythicize the person of Christ and return it into the human dimensions of Jesus-the-man (namely, Bultmann).

We know from the work of the philosopher Carl Jaspers what a deification of man may bring about. We think that his ideas resemble the reason for which Newton, too, could not accept Jesus as God. Perhaps this is another case where Newton was ahead of his times: namely, Jaspers argues that it is truly decisive for the future of Biblical faith, whether it can relinquish Christ-the-God and accept Jesus-the-man:[101]

> We have to give up the religion of Christ that sees God in Jesus; we have to build the event of salvation on the Deutero-Isaiah idea of sacrifice and apply it to Jesus.[102]
> It happens universally that people unreasonably worship a person, promote him to a superman, an ideal of mankind [...]
> No matter what the reasons for the deification of a person, no matter to what degree and in what form, at its root it is always an error.
> A philosophical faith removes the veil from the deified human in any form [...] Right from the start, the deification of one person serves as a means to scorn the rest. [103]

Obviously, Newton's thinking followed a similar path. He always saw in Jesus just a man; nevertheless, he accepted His message as funda-

100 Betty Jo Teeter Dobbs: *The Janus Faces of Genius*, p. 40f. This approach is not unknown in our times: the cosmic interpretation of Christ has its prominent position in the works of Pierre Teilhard de Chardin who, like Newton, neglects the category of self-sacrifice and suffering; instead, he understands Christ physically, as the attraction of the Omega point.

101 Preface by Miloš Havelka in: Karl Jaspers: *Chiffren der Transcendenz* (in Czech: *Šifry transcendence*, Praha: Vyšehrad, 2000), p. 36.

102 *Der philosophischer Glaube*, München: R. Piper&Co. Verlag 1948 (in Czech: *Filosofická víra*, Praha: Oikoymenh, 1994, p. 63).

103 Ibid., p. 83.

mental. He saw the kernel of the message of Jesus in the statement that before God, the highest value is love. One of his works, *Irenicum*, starts with the words:

> In matters of religion the first & great Commandment hath always been: Thou shalt love the Lord thy God with all thy heart & with all thy soul & with all thy mind. And the second is like unto it: Thou shalt love thy neighbour as thyself. On these two hang all the Law & the Prophets. Matt. 22,27.[104]

Newton's opinion evidently came from the correct understanding of the Biblical meaning of faith: obedience and trust in God and His laws. The rest of Europe had to wait until the 20th century Jewish philosopher Martin Buber recognized the difference between an objective and non-objective relationship. So we may declare that Newton's idea was precise, and even far ahead of its time.

Objections to the Divinity of Jesus Christ and to the Trinity

As to the Nicene Council, Newton felt that metaphysical and incomprehensible speculations are contrary to the belief in God and Christ, since the basis for faith is a personal relationship, i.e. the exact opposite of any speculation. Let us quote the beginning of Newton's treatise *Twenty-three queries regarding the word "ομοουσιοσ"*, it was the pivotal term at that Council and means *of the same essence*:

> Quære 1. Whether Christ sent his Apostles to preach Metaphysicks to the unlearned common people & to their wives & children.[105]

Newton even had reservations toward some practices of the Church: e.g., he disagreed with the 27th Article of Faith of the Church of England "De Baptismo," that teaches that the baptism of children shall be kept as equal to the commandment of Christ, (although Christ baptized adult persons only). He wrote:

> After baptism men are to live according to the laws of God & the King & to grow in grace & in the knowledge of our Lord Jesus Christ, by practising what they promised before baptism, & by studying the scriptures & teaching

---

104 *Irenicum*, Keynes Ms. 3 King's College Cambridge, http://www.newtonproject.sussex.ac.uk/.

105 Keynes MS 11: "Twenty three queries regarding the world 'omoousios'."

> one another in meekness & charity without imposing their private opinions or falling out about them.[106]

Such a thing certainly cannot be expected from little children.[107] Incidentally, Newton recognized only three sacraments: baptism, the Lord's Supper, and the laying on of hands.[108] This point again recognizes Newton as a careful reader of the New Testament: Jesus did not perform any other sacraments.

In Newton's times, the Church of England stood in sharp opposition to the Catholic Church. It had its historical reasons – alternating Catholic and Anglican sovereigns. Newton himself maintained strictly anti-Catholic opinions. He certainly did not fail to notice that the Church introduced the worship of people (the saints), of images, of statues and of relics of the saints: in the year 609, Pope Bonifacius permitted the worship of images of the Virgin Mary and of saints as a substitute for the worship of pagan gods.

Newton saw in this a return to pagan idolatry, and a modern Protestant cannot but agree with him. For the worshiping of images and relics stands in grave opposition to the Second Commandment of Protestants (Catholic Commandments omitted it):

> Thou shalt not make unto thee any graven images, or any likeness or any thing that is in heaven above, or that is in the earth beneath, or that is in the water under the earth: thou shalt not bow down thyself to them, nor serve them.[109]

We may find in Jesus' teaching general remarks about replacing God's laws with the commandments of men. Naturally, nowhere do we find any sign of Jesus' worshiping of his mother; quite the opposite: by today's standards, his manner of dealing with her was rather disrespectful.

In the chapter about the Gospel, we have tried to show what Jesus' goal was. We think that, in Newton Jesus had an intelligent disciple who understood the core of his message and accepted a large part of it. (Not all of it, howewer: Newton had never gone beyond the *Mosaic distinction*.) He did comprehend the importance of love, the relationship with God,

---

106 Isaac Newton: *Irenicum*, Keynes Ms. 3, Thesis 19. See http://www.newtonproject.sussex.ac.uk/.

107 Zdeněk Trtík: *Komparativní symbolika*, Praha: Husova československá bohoslovecká fakulta, 1962.

108 Fritz Wagner: *Isaac Newton im Zwielicht zwischen Mythos und Forschung*, München: Verlag Karl Alber, 1976.

109 Exodus 20,2–17, and also Deuteronomium 5,6–21.

the irrationality of speculation, he doubted the deification of Jesus and the lack of justification for the Trinity. Likewise, although formally an obedient member of the Church of England, he privately always tried to hold on to his "keys of knowledge" and never let anybody take them away from him.

In a word, Newton read the Bible closely.

Newton and Moses

Moses established the religion of the Israelites as strict monotheism. God does not have a shape, does not even have a name, He is called *Adonai* = Lord,[110] He is a truly transcendental God. The five books of Moses are the introduction to the Old Testament. Newton repeatedly claimed that the monotheistic religion of ancient Israel was the most rational religion in all of history.

Newton deeply admired Moses. For him, Moses was the union of wisdom and piety, the ideal personification of both a scientist and a priest. Perhaps we may talk of a spiritual affinity of Newton and Moses in their temperance, puritanism, and obedience of the divine laws – the Ten Commandments. Newton did not write a special treatise on Moses, but he frequently mentioned and quoted from him.[111]

Although he never explicitly expressed it, he certainly knew the results of the analysis of his contemporary critics of the Bible, Simon and Hobbs, and accepted them. He admitted that, perhaps, Moses was not the actual author of the entire Pentateuch as it stands in the Bible. He believed that an editor had collected and recorded scattered oral traditions about Moses' acts, thoughts, and opinions, passing the edited text on as if Moses were the author. That, of course, did not change Newton's belief in the authenticity of the Pentateuch and of the person of Moses. The books of Moses, regardless of who penned them, were for him the basic revelation of the Biblical religion, and at the same time the exegesis of all the wisdom of antiquity. They were written in a simple language and could be accepted literally.

Newton believed that Moses had all the wisdom about the created world, but for his uneducated contemporaries he had to simplify it. Thus Newton thought that he himself, through his experiments, was merely removing Moses' narrative encoding of the entire scientific truth. He was

110 Milan Balabán: *Hebrejské člověkosloví*, Praha: Herrmann & synové, 1996, p. 8.

111 Mentions of Moses: *Theological Notebook*, Keynes Ms. 2–18, *Irenicum, or Ecclesiastical Polyty tending to Peace*, Keynes Ms. 3–36, *A Short Schem of <the true> Religion*, Keynes Ms. 7–8, *The Original of Monarchies*, Keynes Ms. 146–10, see www.newtonproject.sussex.ac.uk.

not alone in this attitude toward his own work: several of his contemporaries, Henry Moore, Robert Boyle, and Robert Hooke, saw their own scientific explorations as renewals of ancient philosophy, of the *prisca sapientia*. It was a common opinion in those days that the decay from the original revelation of the *prisca sapientia* occurred in Egypt by sinking into pagan polytheism. Even the term *corrupted knowledge* became fashionable. Moses was believed to have taken the Jews out of Egypt in order to protect them from pagan corruption, reinstate true monotheism, and thus opened the way to true knowledge. Newton was deeply convinced that he himself was likewise merely renewing the lost wisdom.

Newton clearly comprehended the language of the Bible, especially its undeification of nature, this removal of a multiplicity of gods and their actions. For him, *polytheism* was destructive to true knowledge: the recognition of false causes meant that false gods were in mutual conflict, making true knowledge impossible. (Newton placed the *Holy Trinity* in the same class!)

Newton clearly understood how essentially true science depends on monotheism, on the belief that the entire world functions as a whole and not as a combat of various agents (the so-called *constellative action* of several gods) that may be in mutual contradiction. The world must have one creator, so that each of its many parts obeys a single universal order. Such a system cannot exist in a polytheistic world.[112]

Newton's *Principia* was to be a publication of a renewed *prisca sapientia*, the oldest, monotheistically revealed wisdom; and *Principia* carried another theological dimension: it was to secure monotheism for future ages, so that another decline from monotheism should be prevented forever. It therefore seems probable that Newton himself saw in his *Principia* a completely new kind of rational and unquestionable theology, (theophysics), for which he used a heavy weapon: the number = mathematics. And he could also build upon the Bible, on the apocryphal Book of Wisdom.

From a philosophical perspective, we have to consider some religious knowledge: the religion of Moses (or monotheistic religion as such, which shall be discussed later) brings a novelty unknown in antiquity: the distinction between the true God and the false gods. That distinction was unknown in antiquity.

---

112 This topic will be worked out in more detail in the chapter about the *Scholium Generale*. There we shall show that Newton may have received his monotheistic inspiration from non-Biblical sources as well.

Ancient nations respected each other's gods as equivalent and interchangeable, and practiced the kind of translation of their names, according to their areas of competence (for instance, in international treaties). It is the essence of *cosmotheismus*.[113] Thus, for example, the Egyptian god Thoth was translated as the Greek Hermes or Roman Mercurius; all were taken as gods of communication, culture and trade. Likewise, Isis was translated as Aphrodite or Venus, etc. Monotheism stopped that practice, declared its God the only God and other ones false gods, idols. Moses thus carried the distinction between good and evil to its conclusion about God and gods, and alas, started the history of intolerance.

Newton adopted this distinction between good and evil, or correct and incorrect (the Mosaic distinction), for his natural science in its entirety. He thus gave science a fundamental power: since Newton's times, natural science claims quite intolerantly to be the *only* one providing the correct knowledge. Perhaps we may say that within the Mosaic distinction of correct and incorrect, scientists believe that they are the chosen people.

If anybody claims that science is the religion of today, we may add: yes, of course, the ethos of science stands entirely on the Old Testament distinction of good and evil, the correct and the incorrect.

Twelve Articles

This treatise may be interpreted as Newton's theological testament. It presents a sum of how far Newton had arrived in his reasoning: an unorthodox creed in a single God, the human nature of Jesus Christ and the omission of the Holy Ghost.

Isaac Newton's twelve articles on God and Christ[114]

Article 1. There is one God, the Father ever-living, omnipresent, omniscient, omnipotent, the Maker of heaven and earth, and one Mediator between God and Man, the Man Christ Jesus.[115]

Article 2. The Father is the invisible God whom no eye has seen or can see. All other beings are sometimes visible.[116]

113 Jan Assmann: *Monotheismus und Kosmoteismus*, Heidelberg: Universitätverlag C. Winter, 1993.

114 Keynes MS 8, Kings College Cambridge, circa 1710s–1720s, available at http://www.isaacnewton.ca/.

115 I. Timothy 2,5.

116 Compare I. Tim. 6,16.

Article 3. The Father has life in himself and has given the Son to have life in himself.[117]

Article 4. The Father is omniscient and has all knowledge originally in his own breast, and communicates knowledge of future things to Jesus Christ, and no one in heaven or earth or under the earth is worthy to receive knowledge of future things immediately from the Father except the Lamb. And therefore the testimony of Jesus is the Spirit of Prophecy[118] and Jesus is the Word or Prophet of God.

Article 5. The Father is immoveable, no place being capable of becoming emptier or fuller of Him than it is by the eternal necessity of nature. All other beings are moveable from place to place.

Article 6. All the worship (whether of prayer, praise or thanksgiving) which was due to the Father before the coming of Christ is still due to him. Christ did not come to diminish the worship of his Father.

Article 7. Prayers are most prevalent when directed to the Father in the name of the Son.

Article 8. We are to return thanks to the Father alone for creating us and giving us food and raiment and other blessings of this life and whatever we are to thank him for, or desire that He would do for us, we ask of Him immediately in the name of Christ.

Article 9. We do not need to pray to Christ to intercede for us. If we pray to the Father correctly, he will intercede.

Article 10. It is not necessary to salvation to direct our prayers to anyone other than the Father in the name of the Son.

Article 11. To give the name of God to Angels or Kings is not against the first commandment. To give the worship of the God of the Jews to Angels or Kings is against it. The meaning of the commandment is: You shall worship no other Gods but me.[119]

Article 12. To us there is but one God the Father, of whom are all things and we of him, and one Lord Jesus Christ by whom are all things and we by him.[120] That is, we are to worship the Father alone as God Almighty and Jesus alone as the Lord the Messiah, the great King, the Lamb of God who was slain and has redeemed us with his blood and made us kings and priests.[121]

117 Compare John 5,26.

118 Rev. 19,10.

119 Exodus 20,3.

120 I. Corinthian 8,6.

121 Rev. 5,9–10.

Newton the Historian[122]

Newton tried to comprehend the world in its fundamentals. The introductory verses of Genesis (see the chapter on Hexameral literature) were his guide. Just as he had studied physics (the non-living world) and alchemy (the living world), so he tried to study Logos, the domain of the Spirit and meaning, i.e. the area of the interaction of Providence and humanity. And for Newton, that was the domain of history. Frank E. Manuel calls our attention to the fact that Newton elaborated the history of all the nations of antiquity, as well as the entire history of the Christian Church, so that, from his point of view, he covered the human history in its entirety.

He saw in history an endless continuation of theophany, therefore the motto of this chapter might be a quotation from Carl Jaspers: "History is not action in time, but the eternal Incomprehensible that runs across time."[123] From a modern perspective, it seems that Newton was far from being as good a historian as he was a physicist. His historical work was too much under the influence of his religious views.[124] However, that alone is not enough to explain the failure of Newton-the-historian; we shall show later how precisely his physics was likewise a mathematical formulation of his religious confession, something that, however, does not damage the physics.

Even a glimpse of historical conclusions we find somewhat embarrassing:[125] how could a man who invented *calculus* pass such unsubstantiated

---

122 An importand book by Jed Buchwald and Mordechai Feingold *Newton and the origin of civilization* appeared too late to be considered here.

123 Karl Jaspers: *Chiffren der Ttranscendenz* (in Czech: *Šifry transcendence*, Praha: Vyšehrad, 2000, p. 48).

124 Ian P. McGreal (ed.): *Velké postavy západního myšlení*, Praha: Prostor, 1999, p. 285.

125 A brief sample of chronological data according to Newton:

1006. Minos prepares a fleet, clears the Greek seas of Pyrates, and sends Colonies to the Islands of the Greeks, some of which were not inhabited before. Cecrops II. reigns in Attica. Caucon teaches the Mysteries of Ceres in Messene.

1005. Andromeda carried away from Joppa by Perseus. Pandion the brother of Cecrops II. reigns in Attica. Car, the son of Phoroneus, builds a Temple to Ceres.

1002. Sesac reigns in Egypt and adorns Thebes, dedicating it to his father Ammon by the name of No-Ammon or Ammon-No, that is the people or city of Ammon: whence the Greeks called it Diospolis, the city of Jupiter. [...]

994. Ægeus Reigns in Attica.

993. Pelops the son of Tantalus comes into Peloponnesus, marries Hippodamia the granddaughter of Acrisius, takes Ætolia from Ætolus the son of Endymion, and by his riches grows potent.

989. Dædalus and his nephew Talus invent the saw, the turning-lath, the wimble, the chip-ax,

statements? It could not be an intellectual lapse, as was suspected by the early Newtonian biographers. And we believe that we have found Newton's real intention.

Newton tried to introduce law and order into history. We now know that he even tried to find some harmony in historical events and astronomical observations. But we believe that we now have a notion that summarizes all those partial intentions: the *history of memory*.

That notion is introduced by Jan Assmann[126] and it means that history is not studied from the present positivist perspective, but in the way it is remembered, and keeping in mind what is expected from history. At first, it sounds unusual, even fraudulent. However, we feel that such a view may be permissible. We can even document it, for Newton was not the first one to apply it. This principle was applied in a grandiose manner by the nation that Newton adored, the Jews.

It is necessary to add a little explanation: history is never remembered for its own sake.

We sometimes hear that *we are what we remember*. It is the memory that creates the identity of a person and a nation. History as memory carries a cardinal meaning for the present. The internally adopted and remembered past results in narratives: and narratives, or sagas, have a mission. They may become the *driving force of evolution*, (*Motor seiner Entwicklung*) or the *basis of obligation* (*fundierende Verbindlichkeit*). And such narratives are what we call the *mythos*:

> This term is usually contrasted with history; and that contrast brings forth two oppositions: fiction (mythos) against facts (history), and purpose (mythos) against objectivity (history). It is time to do away with such separation. [...] A past that has been accepted as history is a mythos, no matter whether it is fictitious or factual.[127]

Assmann grants the Jewish nation a voluminous chapter, "Israel and the discovery of religion." There he explains how Exodus as a memory,

---

and other instruments of Carpenters and Joyners, and thereby give a beginning to those Arts in Europe. Dædalus also invented the making of Statues with their feet asunder, as if they walked. 988. Minos makes war upon the Athenians, for killing his son Androgeus. Æacus flourishes. [...]
http://www.newtonproject.sussex.ac.uk/view/texts/normalized/THEM00185.

126 Jan Assmann: *Das kulturelle Gedächtnis*, München: *C.H.* Beck Verlag 1997 (in Czech: *Kultura a paměť*, Praha: Prostor, 2001).

127 Jan Assmann: *Das kulturelle Gedächtnis*, München: C. H. Beck, 1997, pp. 75–76. Translated by the author.

not as an actual event, became the basis for a new religion, the religion of Jehovah, that eventually acquired worldwide success, a success due to its retention of memory.

The subject matter of that *memory* was an act of active fabulation, not a record of history as it happened. We now know quite clearly: the story of Exodus does have a factual basis about a departure from Egypt, but includes some important differences. The departure was not experienced by the ancient Hebrews, but by their neighboring tribe, the Hyksos; and those Hyksos were not slaves, but kings ruling Egypt for some hundred years during the 15th dynasty, about 1650–1550 B.C. And the Hyksos did not leave Egypt of their own will, but were expelled as hated foreign rulers. On that point, all ancient sources agree (e.g. Manetho or Josephus Flavius), with the exception of the Bible, which knows nothing about a ruling over Egypt, but describes the sojourn as slavery.

The Hebrews adopted their neighbours' experiences, contradictory to Exodus, reworked them and thus built the *normative historical foundation* of their own nation at the time when their nation needed such a historical support.[128] Although just a fabulation, Exodus became the basis for a new, highly successful religion of worldwide consequences.

Israel was the only nation of antiquity that had survived even the exile. That was an extraordinary achievement, an achievement of *memory*. For all other nations of antiquity, exile was the ultimate tragedy. An inevitable end came through assimilation with the native population. Assmann even coins the term *portable home country* for Israel: it was the all-resistant faith in Jehovah, based on Exodus, which saved the national identity even in hostile environments. Let us quote what Newton says about the retention of identity of the Jewish nation in captivity:

> [I]n a wonderful manner continue numerous & distinct from all other nations: which cannot be said of any other captivated nation whatever, & therefore is the work of providence.[129]

Newton seems to have comprehended the accomplishment of the memory of the Jewish nation and expressed it with almost the same words as did the Hebraist and Egyptologist Assmann three hundred

128 Jan Assmann: *Egypt ve světle teorie kultury,* Praha: Oikoymenh, 1998, pp. 51–71. Original, *Egipto a al luz de una teoría pluralista de la kultura*, Madrid: Akal 1995. Assmann as an Egyptologist can add another proof as support for his thesis: the background of the event of Exodus speaks about Egypt of a later period, not Egypt of the Bronze Age, where Exodus places itself.

129 Manuscript Bodmer MS, 1, f. 8r, also Yahuda MS 7.2g, f. 2r.

years later. The theme of moral appeal is Newton's fundamental motivation: human history turned into mythos binds man by normative requirements, creates a formative force and mobilizes his energies.

Thus we may conclude that, according to our opinion, Newton did not want to write a history in the positivistic sense, but wanted to transform the past into an active, fundamental history, a *mythos*. Newton was labouring in the field of cultural memory. He tried to bring order into the chaos of a myriad of data and act both as a specialist in cultural memory and an *architect of memory*.

His apologetic argument is obvious: he tried to rescue the sense and reason of the past; it was the performance of *rescue-memory*. We believe that Newton wanted to help and make a contribution toward a renewal and solidification of European culture by means of his own interpretation of the past. Obviously, he anticipated that with the emancipation of the sciences and centrifugal atheistic tendencies, Europe may lose part of its memories and identity, thus falling back into chaos. And combating chaos was Newton's perennial challenge.

And, in conclusion, two pearls to think about: Newton took history as a fulfilment of Biblical prophecy. Although he preferred to avoid extrapolating it into the future, he had dared to do it in exceptional cases.

Thus, from the mystical numbers in the book of Daniel, he calculated the beginning of the thousand-year Reign of Christ for the year 2060. In 2003, it caused a shock in the media when the Canadian scholar Steven D. Snobelen mentioned it in his interview with BBC Television. However, the date was misunderstood: Newton spoke about the beginning of the Reign of Christ, while the BBC talked about the end of the world.[130]

And in his book *Observations upon the Apocalypse*, Newton speculates about the second return of the Jews to the Holy Land and the restoration of their state, uprooted by the Romans in 70 A.D. From the favoured numbers 3½, 1260, 1290, 70 etc., Newton derived the following predictions: in the year 1895, the Jews should again reunite in Palestine and their new state should begin in 1944. The accuracy of that numerological prediction is surprising: in 1896, Theodor Herzl published his book, *Der Judenstaat*; and modern Israel dates from the year 1946.[131]

---

130 Steven D. Snobelen: *A time and times and the dividing of time*, see http://www.isaac-newton.org/.

131 Steven D. Snobelen: "The Mystery of Restitution of All Things: Isaac Newton and the return of Jews" in: *Millenarianism and Messianism in Early Modern European Culture: The Millenarian Turn*, ed. J. E. Force – R. H. Popkin, Dordrecht: Kluwer Academic Publisher, 2001, pp. 95–118.

IV. Newton the alchemist

'Tis true without lying, certain most true. That which is below is like that which is above that which is above is like that which is below to do the miracles of one only thing.
Emerald Tablet of Hermes Trismegistos

Nature uses only the longest threads to weave her patterns, so each small piece of her fabric reveals the organization of the entire tapestry.
Richard P. Feynman, Nobel Laureate[132]

Today, the word *alchemy* is a synonym for a mysterious, obscure and somewhat suspect process of fundamental change. We hear about the alchemy of human relations, alchemy of music, we even have a book on economic alchemy. Alchemy in the framework of natural science is viewed as something foolish, futile, at the same time charlatanic or fraudulent. That accusation is not completely unjustified. However, alchemy is also a perennial source of inspiration, because it perceives this world in a different way, thinks in a different way, and registers phenomena for which modern science lacks *sensors*. And we have to admit that the most modern science is approaching ever closer to alchemy. For example, a hundred years ago Ernest Rutherford accomplished the change of one chemical element into another, thus proving that the basic assumption of the alchemists – transmutation of elements – was correct.

The origin of the word alchemy is uncertain. For sure, the *al* comes from Arabic. But for the *-chemy*, we have several mutually contradictory possibilities. Milan Nakonečný, prominent Czech authority on esoteric sciences, claims:

132 Richard Feynman: *The character of Physical law*, London: Penguin, 1992, p. 34.

The origin of the word ALCHEMY is obscure, but may derive from old Egyptian *khemi*, – black, or more precisely Black Earth = Egypt, because alchemy used to be called Egyptian art and ancient Egypt had the reputation of a place where alchemy was widely practiced. It may even be alchemy's cradle. Plutarch derives the word alchemy from Arabic *al-kemia*, which means the same: black earth, Egypt.[133]

Zdeněk Kratochvíl[134] derives the word from Greek χεμεια = juice, René Alleau[135] from Hebrew *Sh-m-sh* = sun, and Jan Assmann[136] from Egyptian *kemit* = perfection, end. Egyptian *chemet* = trade, guild, skill, offers another possibility.

We find the same problem with the definition of alchemy. Professor Nakonečný says:

> In ancient Egypt, alchemy was one of the secret temple sciences that were ascribed to the legendary Hermes Trismegistos, the Greek name of the Egyptian god of culture, Thoth. It was thus also-called Hermetic philosophy. It is incorrect to view alchemy as a primitive chemistry, a historical predecessor of modern chemistry, because alchemy was basically a science of life and its changes.[137]

Zdeněk Neubauer agrees. In his *O kameni mudrců* (*On the Lapis Philosophorum*) he maintains:

> Alchemy is not a predecessor of chemistry. Chemistry studies material changes as if its atoms were a sort of Lego blocks, i.e. something that does not change. That is why chemical reactions can be expressed in terms of equations: the original set equals the final set. Alchemy, on the other hand, explains natural processes as changes in fundamental qualities. It is the very basic matter that changes. [...] An alchemist is thus closer to a modern biologist, particularly an embryologist studying the evolution of a fetus. [...][138]

There are nowadays several books, free of pejorative prejudice, offering factual information on alchemy and its history.[139] A commendable

133 Milan Nakonečný: *Lexikon magie*, Praha: Ivo Železný, 1994, p. 14.
134 Zdeněk Kratochvíl: *Mýtus, filosofie, věda*, Praha: Michal Jůza&Eva Jůzová, 1993, p. 100.
135 René Alleau: *Hermes a dějiny věd*, Praha: Merkuryáš 1995, p. 13.
136 Personal communication via e-mail.
137 Milan Nakonečný: *Lexikon magie*, p. 14.
138 In: Michio Kuschi: *Kámen filosofů*, Bratislava: 1994, Epilogue by Zdeněk Neubauer: *O Kameni mudrců*, pp. 95–111.
139 Helmut Gebelein: *Alchymie, magie hmoty*, Praha: Volvox Globator 1998, original in German:

brief history of alchemy is presented in the recent book *Alchemy* by Brian Cotnoir.[140] He believes that alchemy developed gradually in Egyptian metallurgical temple workshops, documented in the Old Kingdom around the year 3200 B.C. From a much later period, around the third century B.C., we have some recipes for making alloys, and some of them already tell us how to make synthetic metals such as silver. The concept of mythical transformation "from unclean into clean" dates from the earliest metallurgy. Alchemy as we understand it today dates from Alexandria of the 1st century A.D., from the times of syncretism, interplay of various spiritual tendencies and traditions. It absorbed elements of gnosis and Neo-Platonism into its basic concepts (doctrine of salvation through purification), of natural magic (concept of the astral), Greek natural history (four "elements" of Empedocles), Babylonian astrology (quality of time), Christian theology, as well as Egyptian mystery cults of Isis and Osiris. Hand-in-hand with the mixing of those frequently contradictory elements, we find the development of a particular vocabulary, saturated with metaphor, allegory, and symbols.

European alchemy[141] started as a mixture of two currents, both originating in Alexandria. First was Arabic, arriving in Europe as a by-product of the Crusades. Translations into Latin, the language of scholars, came as early as the 12th century. The other came from Alexandria to Greece and Byzantium and after the Turkish conquest of A.D. 1452 it was brought to Florence by educated refugees, together with many literary documents, among them a list of manuscripts known as *Corpus Hermeticum*. Renaissance scholars – we would probably call them magicians – tried to absorb all its wisdom. The Latin translation of *Corpus Hermeticum* was crucial, as will be shown below.

In the books mentioned above – and probably in others as well – we also find "theoretical alchemy," i.e. the three principles of alchemy, (two were enough for Arabic alchemy), four elements, belief in transmutation, and, last but not least, in the production of the obscure *lapis philosophorum*, a *panacea* for all diseases and a substance that may change an

Alchemie, Munich: Diederisch Gelbe Reihe, 2000; Bernhard Dietrich Haage: *Středověká alchymie*, Praha: Vyšehrad 2001, original in German: *Alchemie im Mittelalter*, Duesseldorf: Artemis & Winkler Verlag 1999; Bernard Roger: *Objevování alchymie*, Praha: Malvern, 2005, original in French: Bernad Roger: *A la découverte de l'alchimie: L'Art d'Hermes à travers les contes, légendes...*, St. Jean de Braye (France): Editions Dangles 1999; Claus Priesner, Karin Figala (eds.): *Lexikon alchymie a hermetických věd*, Praha, 2006, original in German: *Alchemie. Lexikon einer hermetischen Wissenschaft*, Munich, C. H. Beck 1998.

140 Brian Cotnoir: *Alchemy*, York Beach (USA, ME): Red Wheel/Weiser LLC, 2006.

141 There also exists a comprehensive, and still alive, alchemy in places such as India and China.

imperfect metal into a perfect one – gold. If the text is long enough, it also includes alchemistic symbolism, (D. Ž. Bor calls them "Renaissance comics") and enigmatic texts, full of clever, mysterious words, unintelligible to the uninitiated.

Sooner or later, the reader feels that those seemingly explanatory texts circumnavigate a complete enigma, miss the substance and try to describe the un-describable. Unfortunately that feeling is justified. In despair, the reader turns to the original: instead of reading books "about alchemy" written by modern scholars, one starts reading books written by alchemists themselves. It is a laborious task: without a thorough preparation, the reader gets hopelessly lost.

Only a few authors among the "non-alchemists" have come close to alchemy. Betty J. T. Dobbs, a prominent authority on Isaac Newton, says: "Alchemy and its non-mechanical processes are a continuation of God's working in the realm of matter."[142]

Zdeněk Kratochvíl writes:

> The origin of alchemy as a meditation of matter must be looked for in ancient Egypt and Sumeria. [...] The laboratory work is a liturgy of copying cosmic events. They unite heavenly and earthly forces. The order of the world is revealed in the laboratory through the life of matter[143] [...] During the Renaissance, alchemy passed through an unusual qualitative expansion and, simultaneously, through a popularization. Alchemy thus became [...] an important part of contemporary culture, [...] its engine and, at the same time, its constraint.[144]

René Alleau[145] reminds us of the opinion of the Arab Gnostic Djabir ibn Hajjan (730–804), known under his Latin name Geber, that every genesis must be followed by an exegesis: a God-created world must be comprehended and explained by man. The way to achieve it is alchemy.[146] The present author found the best inspiration in the work of the famous historian of religion, Mircea Eliade, who studied alchemy as an expression of faith. In his book *Blacksmiths and Alchemists*,[147] he shows that alchemy is a living remnant of a mythical worldview. Briefly, his argument

---

142 Betty Jo Teeter Dobbs: *The Janus faces of genius*, Cambridge, 2002, p. 83.

143 Zdeněk Kratochvíl: *Mýtus, filosofie, věda*, p. 100.

144 Ibid., p. 258.

145 René Alleau: *Hermes a dějiny věd*, p. 27.

146 Geber's idea seems to precede Heidegger by more than a millennium: exegesis (hermeneutics) is not a theoretical entity, but a practical one. It is neither a discovery, nor a wisdom, but a way of existing.

147 Mircea Eliade: *Kováři a alchymisté*, Praha: Argo 2000.

goes as follows: both the ancient metallurgist and the medieval western alchemist see Nature as hierophantic: not only sacred and alive, but also divine. As best as he can, an alchemist cooperates with the Creator and seeks to complete Creation to Perfection. His work is thus sacred. Nature can be changed by means of fire: fire is the principal means of transmutation. Moreover, fire, flame, bright light, even the glare of passion express an important spiritual experience, absorption of the Sacred into one's own, into God's presence.[148] Perfecting Nature means speeding up Time, even abolition of Time in case of Transmutation, therefore liberation of both Nature and the alchemist from the confines of Time.

The alchemists of old had their own way of comprehending the world and their own mission in this world. It was contained in their mythos:

> Mythos [...] is narrated in order to orient oneself in this world and in oneself; it is a truth of a higher order; it not only agrees with reality, it also brings normative requirements and exhibits a formative force.[149]

Mythos and mythical speculation has been yet extensively studied. Vlastimil Rollo[150] presents a clear compendium based on the work of Ernest Cassirer, Carl Gustav Jung, Sigmund Freud and Claude Lévi-Strauss.

The oldest type of thinking, animal realistic thinking, is no longer accessible to us while we are awake. Our spiritual activity approaches it while we are in a state of lower awareness we call dreaming. It is thinking in clear and almost tangible (synesthetic) images. It is an automatic thinking, passive, beyond our power of influence. Our intentions are patently absent. This way of thinking is nowadays contaminated by some rudimentary rationality even among the most primitive peoples. A mixture of an older and younger kind of thinking is what we call mythical thinking. It barely suspects the distinction of oneself from the environment, a distinction clearly recognized only by a fully developed rationality. A fuzzy boundary between an individual and his environment may give rise to magic as an action at a distance, and alchemy as a transmutation of both matter and the alchemist. Mythical thinking lacks all the tools of rational thinking. It knows neither abstraction, causality, logic, nor does it question truthfulness. It is not a means of rational

148 See Matthew 3,11: I indeed baptize you with water unto repentance; but he that cometh after me is mightier than I, [...] he will baptize you with the Holy Ghost and (with) fire.

149 Jan Assmann: *Kultura a paměť*, Praha: Prostor, 2001, p. 70. Translated by the author.

150 Vlastimil Rollo: *Emocionalita a racionalita*, Praha: Sociologické nakladatelství, 1993.

adaptation of man to the world, but one of empathy. We now have two ways of perceiving the world: empathy and thinking. (Earlier, we only had just one, empathy.) The two are mutually independent, with separate mechanisms and even separate physiological processes: limbic system for emotions, neo-cortex for rationality.[151]

Above all, the source of mythos is not a thought, but a feeling. Mythical thinking is controlled by desire and works with a series of synesthetic images directed toward future action through valorisation.[152] Instead of describing the substance of an event and its causality, (the common process of rational thinking), mythical thinking describes the beginning, the source. But in what follows, nothing ever happens by accident. Everything happens due to some purpose, always individually, never as a manifestation of a general principle. The culprit, or culprits (constellation), is always named.[153]

According to mythical thinking, things do not follow through cause and effect, but through a spatial and temporal contact. If two things ever get into a contact, they will never be separated. Since anything may sometime, or somewhere, touch anything else, anything may become anything else. If mythical thinking wants to comprehend a more complicated whole, it often uses a parallel with the human body. This kind of magic anatomy equates parts of the body with parts of the world, so that we find a diffusion of magic anatomy into mythic geography, and back. Such a way of thinking is quite demanding on pictorial memory. It must keep track of all details, while constantly eliciting contributions from empathy.

Mythical thinking cannot analyze, cannot distinguish between a whole and its part. (A tasteless example: veneration of relics of saints.) It is not its fault: it follows from the polytheistic origin of mythos. It has been shown by Assmann[154] that polytheism always has the element of one supreme, hidden, all-penetrating and ever-involved god. But it presents us with a strange relation between a whole and its part. Mythical thinking does not categorize events into classes based on analysis and abstraction. Instead, it uses oppositions, mainly binary ones: up/down, cold/warm, light/darkness, etc., but also triple, quadruple, base seven (planets), base

151 Ibid., p. 13.

152 Valorisation is an opaque process, beyond our conscious influence. It is the true black box of human psyche, endowing a person, item, event etc. with a value, which, of course is valid only for a particular person. This subject will be further treated in this book.

153 Jan Assmann: *Egypt, teologie a zbožnost rané civilizace*, Praha: Oikoymenh, 2002, p. 218.

154 Jan Assmann: *Moses the Egyptian*, Harvard University Press, 1998, p. 73.

twelve etc. Elements of various kinds may be restricted into those *oppositions*. They will then exist in mutual relation through analogy.

Thus number seven will yield *planets = gods = days in the week = metals = plants = animals = precious stones = lands*, etc. Mutual relation of days of the week, e.g. Saturday and Sunday, for a modern man, means no more than the flow of time. But in a mythos, it may have meant the relation, even conflict of gods: Saturnus = Chronos corresponding to Saturday, is defeated by the god of Sunday, Apollo. The same may mean that Bavaria = Saturnus = wolf = lead = Saturday is defeated by Bohemia = Sun = lion = gold = Sunday. Such endless analogies may involve geography, calendar, zoology, botany, colours, anatomy, mythology, astronomy, etc., and thus yield an excess of significance.

Mythical thinking lacks logic, but it has an order. Unlike abstract thinking, it is coherent, never gets into internal conflicts, because its way of reasoning, according to rationality too inaccurate and flexible, is enabled to explain absolutely anything. It also has an element which distinguishes it from abstract thinking: its centre of interest is man, a living, feeling, and suffering being. Mythical thinking does not solve anything but man's emotional and existential well-being. Therefore, it will never be entirely abandoned, particularly by people who have not yet suppressed their emotionality.

Finally Paul Tillich, a prominent Protestant theologian, believes that mythical thinking is always of a religious character. This view agrees with Eliade's statement, that the origins of alchemy are ritual, therefore religious. Alchemy started as a practical application of mythos, of matter's meditation, of a ritual.

Rationality penetrated into alchemy quite slowly, just as self-consciousness and intention slowly creep into our dreams. Psychology calls such dreams *lucid dreams*. Within the aforementioned *archaeology of thinking*, we may say that mankind arrived at its present way of thinking as if it were waking up from a dream.[155] Today's thinking proceeds from a concrete thing to a concept, and then studies the concept. Alchemy, on the other hand, was originally a direct, unspoken interaction, which was rationalized only later. Likewise, perception used to be different. The so-called synesthesis is closely related to mythical thinking: the world seems to be more colourful if we do not think about it. Instead, we just perceive it, feel it. Colours are livelier, sounds are clearer. Our sense of smell, touch and taste play a more serious role.

155 D. Ž. Bor: *Abeceda stvoření.* Praha: Trigon, 1993.

Regrettably, leftovers of mythical thinking may be traced in today's thinking, too. Of course, the two ways of thinking are mutually exclusive. Gradual rationalization of mythical thinking is a positive sign of evolution. Intrusions of mythical thinking into rationality produce intellectual junk.[156] The validity of that statement in the case of alchemy may be demonstrated on text of the *Emerald Tablet of Hermes Trismegistos*, the very essence of the science of alchemy:[157]

One translation, by Isaac Newton, found among his alchemical papers, as reported by B. J. T. Dobbs[158] in modern spelling:

1. Tis true without lying, certain most true.
2. That which is below is like that which is above that which is above is like that which is below to do the miracles of one only thing.
3. And as all things have been arose from one by the mediation of one: so all things have their birth from this one thing by adaptation.
4. The Sun is its father, the moon its mother,
5. the wind hath carried it in its belly, the earth its nurse.
6. The father of all perfection in the whole world is here.
7. Its force or power is entire if it be converted into earth.

7a. Separate thou the earth from the fire, the subtle from the gross sweetly with great industry.

8. It ascends from the earth to the heaven again it descends to the earth and receives the force of things superior and inferior.
9. By this means ye shall have the glory of the whole world thereby all obscurity shall fly from you.
10. Its force is above all force. for it vanquishes every subtle thing and penetrates every solid thing.
11. So was the world created.
12. From this are and do come admirable adaptations whereof the means (Or process) is here in this.
13. Hence I am called Hermes Trismegist, having the three parts of the philosophy of the whole world.
14. That which I have said of the operation of the Sun is accomplished and ended.

---

156 Thus David Hume rejected causality, believing only in some parallelism of events. Likewise, more recently, C. G. Jung and Wolfgang Pauli, Nobel Laureate in physics, offer the theory of synchronicity, another spatiotemporal contact of otherwise unconnected phenomena. See Vlastimil Rollo: *Emocionalita a racionalita*, p. 132.

157 The oldest version may be found in the aforementioned treatise of the Arab alchemist Djabir Ibn Hajjan.

158 Betty Jo Teeter Dobbs: *The Janus Faces of Genius: The Role of Alchemy in Newton's Thought*, Cambridge: Cambridge University Press 1991, p. 274.

The Emerald Tablet is still a beautiful poem. Analysis cannot diminish its beauty. Even today its antique language invites to contemplation. Yes, this text is a paradigm of mythical thinking. It repeatedly makes use of opposition of two, also four. It makes use of mythical anatomy. The author claims to understand all things. He can transform anything into anything. He promises to vanquish the darkness of the world and to grant the glory of victory.

The text dates from the 8th century, so it illustrates the phase when alchemy was already gradually rationalized. But let us remember that from its very beginnings, alchemy was an expression of the original way of thinking, i.e. concrete thinking. We may speculate that, perhaps, in its beginnings, alchemy had no theory and had no concrete goals (!). The definition and enumeration of goals of the later, European alchemy (*lapis philosophorum*, *panacea*, *aurum potabile*, *homunculus*, *palingenesis*[159]) are later principles, sprouting – by necessity – from alchemy. But they obscure the original, demiurgic spirit of alchemy: nature was hierophantic; man's interaction with nature was sacred, ritual, it was the man's meeting of God. That was the ultimate reason and goal of alchemy: the alchemist might partake in a sacred act. There is none, and cannot be any, specification of purpose. This is why alchemical texts are rationally incomprehensible. Nevertheless, their language is comprehensible emotionally, through our feelings.[160]

This opens the question: how could alchemy have contributed to something as different as modern science? Modern science is phenomenological: it knows that it deals with appearances of things, not with things themselves. The representation of environments is not identical with environments themselves. Since the appearances are observed through ways given a priori, more or less to all people, science is valid for all. In a fit of post-Marxist enthusiasm, we may claim that for science, all men are equal.[161] In that sense, science is basically different from alchemy: there the alchemist's very personality is in mythical unity with the matter of

159 See Olaf Rippe et al.: *Paracelsovo lékařství*, Praha: Volvox Globator, 2004.

160 Several aspects of the language of alchemy, its symbols, metaphors, overall leaning toward images, phonetic kabbalah, the perspective of energy, and converting power of that language cannot be treated in this book.

161 Science is meant to be taken phenomenologically: following principles of rational thinking. Alchemy follows principles of mythical thinking; however, that does not mean that alchemy could not be aware of phenomena of its environment. In India, we have evidence for a rational comprehension of the environment. Indian "maya" corresponds to our "appearance of the world." See Usrhs-Pha: *Úvod do metafyziky a esoteriky Indie*, Plzeň: Trigon, 1995.

his experiment. We may say that there will be as many results of a given process as there are alchemists.

Natural science is a consequence of man's self-recognition, which opened up the gateway to rationality. And rationality opened the question of (objective) truth. The step from mythos to rationality also brought about the schism of religion and science: early mythos included them both. Let us briefly look at the parallels between religion and science. Jan Assmann believes that this psychological step led to the sudden appearance of monotheism. He sees in it such a fundamental, revolutionary, inexplicable transition from previous polytheism that he calls monotheism, a counter-religion (*Gegenreligion*). What matters is not the number of gods, (i.e. just one), but God's claim to be the only true one, that this faith is the only right one, is different than the other faiths, (actually: it stands opposite to the other ones), stands on negation, limitation and intolerance. According to Assmann, science acts likewise: it is negative and intolerant. (No wonder: science originated in the lands of monotheism.) He says:

> Scientific knowledge is *anti-scientific* (*Gegenwissen*), because it knows what is incongruent with its theses. Only such an *antiscience* may create rules of what is and what is not true knowledge, i.e. what is a second rate knowledge. Such a method corresponds to a *second rate religion* (*polytheism*) as seen by a monotheist.[162]

Natural science distinguishes between subject and object, which would be impossible in mythical thinking.[163] Science categorizes our experiences (not the things themselves). It thus derives rules universally valid for our perception of the world. As a consequence, the flood of new knowledge replaces the old one.

But something far worse is going on: "Increasing scientific knowledge enlarges ignorance in fundamental matters."[164] Accumulation of knowledge easily obscures many fundamental differences. For example: natural science intentionally omits anything that is not common to all people, such as values. Accepted values fully exhibit differences among various

162 Jan Assmann: *Die Mosaische Untescheidung*, München: C. H. Beck, 2003, p. 24.

163 Although it is not immediately evident, terms subject and object derive from two different Latin words: object, *ob-iacio*, to throw forward or under, and subject, *sub-iaceo,* to lie down. Of course, the roots of infinitives are in both cases the same: sub-icere and ob-icere.

164 Karl Jaspers: *Der philosophischer Glaube*, München: R. Piper&Co. Verlag, 1948 (in Czech: *Filosofická víra,* Praha: Oikoymenh, 1994, p. 38).

people and their individual emotional dispositions. Science intentionally neglects those differences; otherwise it would lose its universal validity. But that reveals the consequence: science cannot deal with questions where values matter. And it is values that matter in this life and in this world, for it is our emotions that make sense for us and shape our decisions.

Alchemy presented a certain – perhaps insignificant[165] – contribution to modern chemistry and biology; nevertheless, it never ceased significantly contributing to modern science. It keeps supplying mainly its sensibility and what Assmann calls "normative requirements and formative powers" of mythos: not just a belief in a possibility of change of Nature, but mainly the faith in unlimited progress. As Eliade maintained:

> Most of the heritage (of alchemy) may be found ... in the ideas of naturalists, in the capitalistic system, in liberal and Marxist economic policies, in the secularized theologies of materialism, positivism and of endless progress, in the overall worship of the unlimited opportunities of homo sapiens, in the eschatological meaning of human labour, technology and scientific exploitation of nature.[166]

Certainly, both nature and human labour had in the meantime lost their sacredness. We no longer have the liturgical dimension that made even the hardest labour bearable.

As mentioned above, it was the Renaissance and its protagonists who gave us the modern scientist, a magician-scholar balancing on the knife-edge between Christianity and hermetic philosophy, the philosophy of the *Corpus Hermeticum*. The example par excellence of such a Renaissance scholar is Giordano Bruno who believed that this *Corpus* was an authentic source of the Bible, the Greek philosophy, as well as of some of the original, cosmic religion. *Hen to pan.*[167] All is One. One divine Being, one universal principle of existence. Here we find the foundations of our science: in the idea that a Source yields a Unity of Everything, being omnipresent, moving everything, and, furthermore, every individual thing or event is merely an independent example of a universal, hidden Deity – and here we see a clear idea of a "universal principle," cosmic

165 This statement requires further research. There seems to be just a small amount of alchemistic wisdom applicable for the budding natural science. Only recently has science recognized and confirmed some of the intuitions of ancient alchemists.

166 Mircea Eliade: *Kováři a alchymisté*, p. 133. Translated by the author.

167 From the Christian point of view, this was pantheism, therefore ultimate heresy and Bruno's death was its natural consequence.

natural law, (not at all a self-evident, trivial idea!) that all processes are always manifestations of one universal Law, one God. And alchemy – itself naturally pantheistic,[168] otherwise it would lose its meaning – applies that idea practically, experimentally. If you would forgive the formulation, from our point of view, it only seems to be an experiment; it is still a ritual, it is working with a sacred living matter of this world. Alchemy follows Nature's footprints; it imitates Nature and asks it questions. Those *imitatio, inscenatio et proiectio* our science borrowed from alchemy. Repeated experiments produced equivalent results and thus gave us the "principle of verifiability." It started as a magic trick of the Renaissance magician: demonstration of a phenomenon via imitation, *imitatio*.[169]

Until the time of Newton, the Renaissance had its sincere, "alchemistic" goals: gradual perfection of Nature and, hand in hand, perfection of man, for the two could not be separated. The goal of the scholar – either Paracelsus, Dee, Andreae, Fludd, Comenius, or Newton – was to study Nature and, simultaneously, improving man, his knowledge and his education. Those men tried to establish a method, which would integrate non-sectarian Christianity, hermetic philosophy and modern science. They still looked to alchemy as "an ambitious endeavour: perfection of man through a new educational method."[170] With all their vigour, they tried to improve man's education in this way, even the general conditions. They sought the New Era.

The New Era did materialize; however, it lost the spirit of reform, or, at best, included it only partially. With the magnificent rise of rationality, the balance of reason and sentiment vanished. The scholars mentioned above expected a different type of scientific revolution: perhaps for the first and last time in the history of mankind, they possessed a balance between reason and sentiment.[171] Within themselves, those two could unite rational thinking with sensual, mythical, alchemistic approach. They could balance them and mutually correct them.

Naturally, that unity could not last. Alchemy postulated the unity of the alchemist and his environment. Modern science distinguished the subject from the object and the old symbiosis had to go. Man as such was pushed aside: science "does not see" and "cannot see" the person,

---

168 Or at least panentheistic.

169 Zdeněk Neubauer: "Corpus Hermeticum Scientiae," in: *Logos, sborník pro esoterní chápání života a kultury*, Praha: Trigon, 1997, pp. 8–20.

170 Mircea Eliade: *Dějiny náboženského myšlení III*, Praha: Oikoymenh, 1997, p. 252. See also Betty Jo Teeter Dobbs: *The Foundations of Newton's Alchemy*.

171 Vlastimil Rollo: *Emocionalita*, p. 153.

because the person is not, nor can they be, a mere phenomenon and object. But the mythos of endless progress through scientific experiment ascended its road to triumph.

Eliade in his papers repeatedly claims that science, through its improvements of nature, inherited alchemy's idea of speeding up, even overcoming time. The present author feels that the same has happened to spatial distances. Science tries its best to enlarge the minute and reduce the large. Science thus seems to attack the a priori principles that make science what it is. The deeper science penetrates into the mysteries of the cosmos or the minute constituents of matter, the less valid is our "reason," and, eventually, reason loses its validity altogether.

Observations of the structure of the atomic nucleus include the observer. However, we would thus return to the gloriously overcome mythical participation of man in his environment. It would be *non-scientific*, *pseudoscientific*, and even *alchemistic*! Science seems to have outgrown not just its Renaissance creators: it seems to have outgrown itself, when it violates its own principles. Science seems to have forgotten that it does not work with things, but with appearances of things, and that there is no place for the individual. As mentioned above, there is no way back.

Intrusion of mythos kills rationality. Alchemy may have been the cradle of science. But it would definitely be its grave.

Let us quote Jaspers' sentence once again, this time in its entirety:

> Increasing scientific knowledge enlarges ignorance in fundamental matters and thus identifies the limits which become meaningful from another, non--scientific perspective.[172]

Newton as an alchemist of his times

Newton's work in the field of alchemy is unique in modern times.[173] It consists of extensive excerpts from the works of other authors, sometimes even their literal copies, and of his own laboratory protocols, as well as theoretic alchemical speculations. The manuscripts contain some 1,200,000 words. There is consensus among Newtonian scholars that his

172 Karl Jaspers: *Der philosophischer Glaube*, München: R. Piper&Co. Verlag 1948. (In Czech: *Filosofická víra*, Praha: Oikoymenh 1994, p. 38. Translated by the author.)

173 Chapters about Newton the theologian used to start in a similar way. Although he is one of the originators of natural science, his most voluminous and significant contributions are in the field of theology and alchemy.

most productive period was between the years 1669 and 1696. At that time, Newton worked at the University of Cambridge, where he lived on the first floor in the Great Court of the Trinity College. His laboratory was in a small wooden cottage. According to his assistant, Humprey Newton, he worked mainly in spring and autumn, when the furnaces were burning for weeks on end. His most intensive period of alchemical work culminated with the new era of his life, when alchemy became Newton's official employment.[174] As an alchemist Newton came a long, long way. He reverted to a young boy experimenting in the laboratory of his house-keeper, a pharmacist, and ended up as the Warden of the Royal Mint. No doubt that was the place where he could put his alchemical experience to practical advantage.

Newton shared the basic alchemical ideas of his times. Like his contemporaries, he wanted to become a *philosopher of fire*. Nevertheless, several of those alchemists became founders of modern scientific disciplines, too. For example, Robert Boyle, while an avid alchemist, is a pivotal figure in modern chemistry. Incidentally, for several years, Boyle worked closely with George Starkey, who influenced Newton greatly.[175]

Alchemists in those times distinguished between common or *vulgar* chemistry, approximately corresponding to today's inorganic chemistry, and *vegetation chemistry*, exploring processes of general vegetation, not only of plants and animals, but of planets, metals and minerals, too. This *vegetation chemistry* is correspondent to modern organic chemistry and biology. Those latter processes were believed to be controlled by the "vegetable spirit," a direct mediator from God, permeating all of nature:

> Nature's action are either vegetable or purely mechanical [...] an exceedingly subtile & unimaginable small portion of matter diffused through the masse which if it were separated there would remain but a dead & inactive earth [...][176]

---

174 We do not list here the manuscripts from his "Mint period." They are too numerous, (c. 900 items) and are accessible at http://www.newtonproject.sussex.ac.uk/prism.php?id=48.

175 George Starkey, who later used several pseudonyms, mainly Eirenaeus Philalethes, was born in the Bahamas. Among the first graduates of the first American university, Harvard, he moved to London where he held various employments and cooperated with Robert Boyle. As a physician, he worked most selflessly during the plague epidemic, contracted the disease and died. According to some scholars, he is the first American scientist of consequence during the "scientific revolution." For more information, see William R. Newman: *Gehennical Fire, The Lives of George Starkey*, Chicago: The University of Chicago Press, 2003.

176 Manuscript *Of Natures Laws*, Dibner Collection MSS 1031B, after Betty Jo Teeter Dobbs: *The Janus Faces of Genius*, Cambridge: Cambridge University Press, 2002, p. 30.

For this spirit of life, Newton uses the name *Magnesia*, according to the alchemist Michael Sendivogius, who introduced this term in his *Novum Lumen Chymicum*.[177] There is a primeval alchemical idea of cosmic spirit of life and love, the spirit of the mysterious harmony of all things, *Harmonia Mundi*, and reflections on that spirit occupied Newton for decades. It is possible that this "spirit" had given him the idea of a universal gravitation, the mutual attraction of all material particles. The possible connections of Newton's alchemical research and discoveries in the field of physics have been thoroughly examined by the American scholar, B. J. T. Dobbs.[178]

Although Newton denied it, from time to time he had tried to find the true nature of gravitation. Perhaps he was inspired by his ideas on the alchemical "magnet." His speculations about the spirit of life might have also inspired his work on optics, because, for a very long time, he thought that the spirit of life may be the essence of light.[179]

Newton's work in the field of alchemy also included a study of internal structure of matter itself.[180] He was inspired by the aforementioned George Starkey, who believed that matter consisted of complex particles, which may be broken down into smaller particles, until eventually we would arrive to the smallest elementary constituents of matter. However, these smallest particles are not yet homogeneous: they are thicker and heavier inside, and lighter and thinner closer to the surface. There is a striking similarity with the modern model of the atom with its heavy nucleus and lighter "shells." These ideas Newton then applied in his studies of light and its interaction with surfaces of materials.

It is interesting how Starkey's and Newton's ideas about the internal constitution of matter lead toward the ultimate goal of alchemy, the *transmutation*. They believed that dividing matter into the tiniest particles was the first step toward transmutation. Those particles may then

177 Zbygniew Szydło: *Water Which Does Not Wet Hands*, Warszawa: Polish Academy of Sciences, 1994.

178 Betty Jo Teeter Dobbs: *The Janus Faces of Genius*, pp. 89–256.

179 Betty Jo Teeter Dobbs: *The Janus Faces of Genius*, pp. 37–46. Dobbs says literally: "That the vegetable spirit might be identified with light is implied by a suggestive term used by the alchemists – illumination."

180 Betty Jo Teeter Dobbs: "Newton's Alchemy and His Theory of Matter," in *Isis* LXXIII (1982), pp. 511–528; Karin Figala: "Newton as Alchemist," in *History of Science* XV (1977), pp. 102–137; Karin Figala: "Die exakte Alchemie von Isaac Newton," in *Verhandlungen der naturforschenden Gesellschaft in Basel*, Band 92, pp. 157–227; Arnold Thackray: *Atoms and Powers. An Essay on Newtonian Matter-Theory and the Development of Chemistry*, Cambridge: Harvard University Press, 1970.

be rearranged according to a new formula and thus the new substance is made up.

Newton had progressed much further than his predecessors. He managed to quantify their qualitative discoveries, and tried to express the relationship between particle size and stability. The opposite of alchemical notion of stability, is the notion of so-called *volatility*, and it is one of key alchemical terms. For Newton, as for all the alchemists of his times, reaching volatility of a substance was the signal that a substance had been divided into its smallest constituents and is therefore ready for the next step.

The next step was to rearrange, re-configure the substance. And that was to be achieved by means of the *philosopher's stone*. That stone was the carrier of both the desired *informatory formula,* and the needed energy for this process. In the words of the alchemist, it provided adequate *fermentative power*.

Finally, let us say a few words about the *opus magnum* of alchemy, producing the *philosopher's stone*. Newton worked on that task for a long time. He followed the so-called *short* or *dry* path, using antimony. There was also a *long* or *wet* path, using lead. It was probably most thoroughly developed by Sir George Ripley, an Englishman whose books Newton also owned.[181] Ripley's method is still used by several of today's alchemists, but it had never been considered a favourite: it was slow and, moreover, *too safe*. Alchemy always carried along the danger of risking one's life (!), it had always been a real existential life-challenge.

Newton followed the *dry path*, because he had full confidence in his own laboratory craftsmanship. He also may have followed the *Triumph-Wagen Antimonii*[182] by Basilio Valentin, and some scholars suppose[183] he might have come after the work of Elias Ashmole[184] and Paracelsus. But he undoubtedly mainly followed George Starkey.[185] The method of

181 According to Harrison: *The library of Isaac Newton*, p. 227:
405 Ripley George: *Opera omnia chemica, quotquot hactenus visa sunt*, Casselis, 1649 *Notes by Newton,* references to alchemical books, mainly on pp. 18, 101, 123, 171, 383, 401, several signs of dog-earing.

182 Basilius Valentinus: *Triumph-Wagen Antimonii*, Leipsig, 1604.

183 See, e.g., Frances Yates: *The Rosicrucian Enlightenment*, Praha: Pragma, 2000, pp. 223–231.

184 According to Harrison: *The library of Isaac Newton*, p. 91:
93 Ashmole Elias: *Theatrum chemicum Britannicum*, London, 1652.

185 According to Harrison: *The library of Isaac Newton*:
554 Æyrenæus Philalethes [i.e. G. Starkey]: *Ennaratio methodica Trium Gebri medicinarum, in quibus continetur Lapidis philosophici vera confectio.*London, 1687.
838 Anonymo Philaletha [i.e. G. Starkey]: *Introitus apertus ad occlusum regis palatium*, Amstelodami: 1667.

making the *Philosopher's stone* is at present known at least in principle:[186] its first step, the preparation of the *Star Regulus of antimony* has been tested at the University of Indiana, USA.[187]

Summary of chapter IV

In conclusion, let us look at alchemy as a precursor of modern science. Its ancient ways of thinking make it difficult to be comprehended; on the other hand, that way of *mythical* thinking was followed by Newton simultaneously with his scientific work, therefore it could help us discover the rationale and the marginal conditions for the origins of modern science.

---

961 [by G. Starkey]: *Liquor Alcahest, or A discourse of that immortal dissolvent of Paracelsus & Helmont*, London, 1657.
1034 Eirenæus Philoponos Philalethes [i.e. G. Starkey]: *The Marrow of alchemy, being an experimental treatise, discovering the secret and most hidden mystery of the philosophers elixir*, London, 1654.
- *The marrow of alchemy* (by Eirenæus Philoponos Philalethes [i.e. G. Starkey] 1709, see 1644)
1296 Eyreneus Philoctetes [i.e. G. Starkey]: *Philadelphia, or Brotherly love to the studious in the hermetick art*, London, 1694.
1300 Phyloctetes, Eyreneus, pseud. See Starkey, George.
1407 Eirenæus Philalethes [i.e. G. Starkey]: *Ripley reviv'd, An exposition upon Sir George Ripley's hermetico-poetical works*, London, (1677)–1678.
1478 Eyræneus Philaletha Cosmopolita [i.e. G. Starkey]: *Secrets reveal'd: or An open entrance to the shut –palace of the King*, London, 1669.
1553 Starkey, George: *Pyrothechny asserted and illustrated, to be the surest and safest means for art's triumph over natures' infirmities*, London, 1658.
1644 Eirenæus Philoponos Philalethes [i.e. G. Starkey]: *A true light of alchymy. Containing, I. A correct edition of the Marrow of alchymy*, London, 1709.
In Harrison's book published in 1978 there are several statements to the effect that Starkey used a pseudonym, particularly that of Eyreneus Philalethes, therefore it is not a recent discovery. Starkey is the author of the largest number of books in Newton's library, and he is also the most frequently quoted author in Newton's alchemical manuscripts. According to Richard Westfall (quoted in William R. Newman: *Gehennical Fire, The lives of George Starkey*, Chicago: The University of Chicago Press, 2003, p. 229), Newton quotes Starkey – Philalethes three hundred and two times.

186 Alchemists searched for support in other mystical systems and the Hebrew Kabbalah seemed to offer itself. The "tree sefirot" comes from the book *Oidipus Aegyptiacus* by Athanasius Kircher. The pictures represent the parallels among systems of planets, metals, animals, plants, etc. and justify the progress from the lower sefirot *Malchut* (Kingdom) toward sefirot *Tifferet* (Beauty). In the system of metals, it corresponded to the step from antimony to gold.
Incidentally, if we look at Mendeleev's periodic table of chemical elements, we notice that gold and mercury are in adjacent places. Therefore, we "only" have to change one proton in the nucleus of mercury into a neutron and mercury will turn into an isotope of gold: ${}_{80}^{197}\mathrm{Hg} \rightarrow {}_{79}^{197}\mathrm{Au}$.

187 See http://webapp1.dlib.indiana.edu/newton/index.jsp.

Alchemy probably provided the inspiration for the scientific method: it always depended on experiment, therefore on an imitation of natural processes. Newton himself was a skilled experimenter in all disciplines and used experimental results as a counterweight to pure speculation.

Alchemy was the cradle of several sciences because it offered hope for changing nature, hope for a limitless progress, hope for discovery, hope for the freedom of learning, hope for overall better conditions thanks to learning, education, etc.

As far as the way of thinking is concerned, we may say:

Alchemical way of thinking is complementary to the rational thinking. Rational thinking generalizes, simplifies, categorizes and qualifies, but it neglects qualities and values. It further pertains to investigations of a large number of events,[188] therefore it tends to disregard exceptions.

Alchemy, in contrast, keeps quality in mind and actually builds on unique events and properties.[189] Everyday observations, after all, pertain to a particular person and a particular event. Those uniquenesses count and have their consequences: for example, a family consists of unique people who cannot be replaced, nor substituted by others. In science, we all may be regarded as people: but in life, we are all different and have different life histories. We may look alike in some respects; but in other respects, what pertains to us does not pertain to anyone else. Our own experience may control our mental stability[190] and cannot be abandoned. It remains in the realm of alchemical, qualitative and value thinking.[191]

Alchemy is not metaphysical, as long as we understand metaphysics in its traditional, static, sense. For metaphysics, the world does not change;

---

188 Richard Feynman: *The Character of physical law*, p. 35: "In thinking out the applications of mathematics and physics, it is perfectly natural that the mathematics will be useful when large numbers are involved in complex situations. In biology, for example, the action of virus on a bacterium is unmathematical."

189 This is why alchemy, as mythical thinking, fails in categorizing: when we like something, we do not care whether it is a flower, smell, animal or a piece of music. It is just a source of pleasure. Therefore, rational thinking fails here completely.

190 The ideal of scientific progress is "objectivity free of prejudice." Ad absurdum, it would require a scientist without values, therefore a psychopath without any ethos.

191 In this connection, we may even speculate about a *de-construction of memory*, similar to that in the relation between Egypt and Israel. See Assmann, Jan: *Moses der Ägypter*, Frankfurt am Main: Fischer Taschenbuch Verlag, 2004, p. 25. Here it is Alchemy vs. Science, too: Egypt was an archaic, outmoded, idolatrous decay, while in alchemy, it was superstitions, ignorance and inaccuracies.
However, such a "de-construction" becomes a cradle, a starting point of everything new and progressive. The most ancient memory must be overcome, but it is still present. Later rational thinking derives from alchemy and any thinking in values remains there.

it remains in its permanent metaphysical condition. Alchemy is dynamic, the world is alive, ever-changing, controlled by *physis*, therefore full of life, full of God. Alchemy studies the changes and wants to take part in them. That is why alchemy builds upon experiments. They not only imitate nature: they raise questions, and they are supposed to keep us informed about the changes and their regularities.[192] These dynamic processes of growth and structural change take place in the field of quality and value. The experimenter is personally engaged in his experiment, he is informed by the experiment; therefore he is literally *informed*, i.e. transformed. This is Plato's "fifth level," an interconnection of learning, knowledge and transformation of being. This is why alchemists of Newton's time also sought the improvement of public matters (Comenius): a change in the quality of contemporary existence.

Alchemical thinking involves the interaction of two subjects: the alchemist works with the living, divine matter; he confronts God, the divine *You*; he works with an absolute value. Perhaps alchemy thus anticipated modern personalistic philosophy, Martin Buber and his non-objective thinking.[193]

If the experimenter meets God in matter, this God has to be substantially present in the matter. But that is impossible for a transcendental God of the Bible. The God of the alchemist must be an immanent God of *pantheism* or *panentheism*.

192 It follows from the idea that God, besides many other roles, also acts as a gigantic "cosmic memory." Some scholars speculate along those lines: e.g., Rupert Sheldrake and David Bohme.

193 Martin Buber: *Ich und Du,* Gerlingen: Schneider, 1994.

V. Newton - the natural philosopher (scientist)

> *The next question was – what makes planets go around the sun? At the time of Kepler some people answered this problem by saying that there were angels behind them beating their wings and pushing the planets around an orbit. As you will see, the answer is not very far from the truth. The only difference is that the angels sit in a different direction and their wings push inwards...In other words, although I have stated the mathematical law, I have given no clue about the mechanism [...] There is always an edge of mystery, always a place where we have some fiddling around to do yet.*
> Richard P. Feynman, Nobel Laureate[194]

Newton's pivotal scientific works:
Philosophiae Naturalis Principia Mathematica – 1687, 1713, 1726, always in Latin.
Opticks – 1704 in English, 1706 in Latin, 1718 in English (posthumous 1730).

Newton's contribution to modern science is usually described as significant, if not fundamental, in three areas – mechanics, optics and mathematics. That part of Newton's work[195] has been thoroughly studied and is well known. Some of it has been a part of the syllabus of physics courses to this day throughout the world. We shall therefore mention it only there, where the context requires it. Our own interest will focus on Newton's trail of thoughts when he worked his way toward his discoveries of fundamental natural principles, his logics, chronology of ideas, their system and, of course, Newton's methodology, which prevails in natural science to this day.

194 Richard P. Feynman: *The character of Physical law,* London: Penguin, 1992, p. 18 and 33.

195 This part of Newton's work is often referred to as science; actually, it is an anachronism, Newton himself called it *natural philosophy*.

Newton was the one who finally convinced the educated public that not all of Aristotle's ideas were correct: a falling body is not "looking for its natural place," nor is "matter itself [that] which is the cause of individualization of bodies and their shapes." There was a growing scepticism in Newton's time about the reliability of our knowledge: the Renaissance reached a crisis and had to seek new epistemological reliability. Can we build our knowledge on mathematics and speculation (René Descartes), or on the examination of phenomena (Galileo Galilei), or can we derive all our knowledge from Biblical prophecies and their interpretation (John Dury)? Newton's solution may be called a *coniunctio oppositorum*, a combination of opposites in the broadest sense. We may even call Newton a modern syncretic. He took into account everything that was known in his time and searched for a meaningful harmony. To claim that Newton built his science on mathematics, experiment, observation and induction is still a simplification that does not encompass his contribution in its wholeness.

Newton tried to discover the divine principles, hidden behind the veil of nature and history. He wanted to observe God in action by studying both nature and history. To achieve this objective, he tried to balance and correct all various approaches. Due to his extraordinary range of knowledge, it was necessary to build not only on his direct predecessors in the field of physics and mathematics,[196] but also to exploit the methodological potential of the Bible, patristic writings, as well as a variety of important ancient philosophical schools – Pythagoreans, stoics, Neo-Platonists, atomists, and also of *Prisca Sapientia*, namely Hermeticism, alchemy, etc. McGuire wrote about Newton:

> Moreover, he believed that the more ancient sources, including those of alchemy, were less likely to be corrupted. For these reasons they were probably closer to the original wisdom in religion and natural philosophy. Accordingly, Newton looked to antiquities as sources of Divine Wisdom for clues pertaining to the structure of the cosmos [...][197]

196 In physics, it was the sequence: Nicolas Copernicus – heliocentric system; Giordano Bruno – infinite universe; Galileo Galilei and Francesco Patrizi – a vision of harmony; Tycho Brahe and Johannes Kepler – interconnection of experiment and intuition; Roberval, Wallis, Cavalieri, Peregrin, Bradwardine, and Oxford mathematicians – mathematics.

197 *Prisca sapientia*, i.e. the *Corpus Hermeticum*, were believed to be a common source of both the Bible and Greek Philosophy. And Newton believed that the older the sources, the less corrupted they were. He believed the same about the alchemistic sources. They should therefore be closer to the original wisdom, both theological and scientific, and Newton studied them as a source of wisdom, because they held the key to the structure of the cosmos. McGuire,

There were minor or major disagreements among those schools. For example, in the field of mathematics, Newton tried to reconcile two contradictory systems: the Pythagorean School was established on the idea of harmony of the world and sensory evidence. On the other hand, there was the Arabic *algebra* with a formal operation without sensory evidence, but embodying the idea of a fateful necessity, *kismet*. Newton's method of self-correction, balancing of the mind, senses, and God's revelation, is a meeting place of both scientific and extra-scientific components. It needed further examination. Alchemy operated on mythical thinking, while the Old Testament prophets and hermetic writings required a sapiential approach. Many of Newton's contemporaries, such as Descartes, were already thinking rationally. Newton was familiar with all those methods and used them as mutually equivalent.

The prevalent view nowadays is that of Professor Dobbs, maintaining that it was Newton's method of self-criticism that helped him gain superiority over his contemporaries and brought about his permanent reputation:[198] he synthesized the discrepancies among various disciplines into a rational unity. Thus he could not only describe the world, but also, up to a point, predict it. This helped him to answer the question "how?" Nevertheless, he refused to explain the natural phenomena in terms of causes and effects. He refused to answer the question "why?" Since his time, this has been one of the principal rules of modern science.

This process of balancing out various approaches to a given problem is a complicated one; we can only speculate about its complexity. However, we may demonstrate it on some particular examples.

Thus Dobbs followed the progress[199] of Newton's idea of universal *gravitation:* he conducted his alchemical experiments and at the same time wrote his *Principia*. She describes his slow and gradual struggle toward the result which, today, may look deceptively simple.

At first, Newton speculated that gravity is propagated by the hypothetical *ether*. Around 1675, he had the idea that the ether's spirit, which also possessed a power to give life, condenses toward the earth's surface. At that time, he was thinking only about terrestrial gravity, in other words, about weight, (in Latin *gravitas*, where the word *gravity* comes

---

James E.: "The Fate of the Date: The Theology of Newton's *Principia* Revisited," in: *Rethinking the Scientific Revolution*, Margaret J. Osler (ed.), Cambridge: Cambridge University Press, 2000, p. 284.

198 Betty Jo Teeter Dobbs: *The Janus faces of Genius*, Cambridge: Cambridge University Press, 2002, pp. 8–12.

199 Ibid., pp. 19–255.

from). Only later did he extend his idea upon the Solar system; and, after 1680, upon the entire universe. Moreover, he realized that it was a mutual attraction between any two bodies. That was the deciding novelty which promoted gravitation to a universal cosmic force.

The research of Newton's works today is rather complicated: his works are monothematic. His alchemistic papers do not make us suspect that the author was a scientist. His scientific books do not indicate the author's historical interests, neither do the historical ones reveal the mind of an alchemist. Conceptual transfers are not obvious. That practice has been adopted by modern science: its assumptions are not stated explicitly. Science as such is actually a multitude of separate fields. Only in recent decades have their overlaps been explored.

Newton divided his world into separate domains according to the required methods of exploration. That division was inspired by the first chapter of the Bible, Genesis 1,1–27.[200] On the first day of creation, God created the non-living matter: therefore, Newton used *physics* in order to study the lifeless matter of Domain No. 1. (Incidentally, in alchemical terms this area is called *caput mortuum*, literally *head of death*, and Newton was very well aware of it, no doubt.) It comprised mechanics and optics, and for its exploration, he developed his method of mathematical physics. That method, applied to that domain, has worked remarkably well. However, let us not forget that it was a method tailored for non-living matter.

Newton had certainly not intended to apply that method to other areas,[201] such as his Domains 2 and 3. He studied those domains much more than his Domain 1, but used a different method. For example, when he studied living matter in Domain 2, he applied the method of alchemy. Many older authors, and some recent ones as well, blamed Newton that his world is lifeless,[202] mechanical, like clockwork. It is a misunderstanding: Newton was fully aware that there are areas where his mechanistic method is not applicable. This is one of the points where modern scientists neglected insight of one of the founders of science: they tried to apply the method tailored for Domain 1 also for other domains.[203]

We shall demonstrate the full power of Newton's thinking on two examples taken from optics and mathematics.

---

200 See Chapter II.

201 Domain 2 encompasses living matter; and Domain 3, that of *logos*, in harmony with Genesis 1,1–27; see Tables in chapter "Hexameral Literature."

202 Morris Berman: *The reenchantement of the World*, Ithaca: Cornell University Press 1981.

203 Thus it can happen that *biology*, or *life science*, still lacks a definition of *life*; and *psychology* even denies the existence of a *psyche* = soul, a part of its very name!

We have already mentioned that in his study of the constitution of matter, Newton followed his older contemporary, George Starkey. Among his interests were the interaction of light and the surface of bodies. At the time of its publication, his *Opticks* was more influential than his *Principia*. Only later was the situation reversed: *Principia* are valid permanently, whereas his *Opticks* is now only of historical interest, because Newton was unable to provide a mathematical formulation for several optical phenomena.[204]

Newton is credited with the discovery that white light is a composition of all spectral colours. But we seldom appreciate the meaning of that discovery. Newton studied (but did not solve) the question: what it means that different bodies have different colours? Nevertheless, he realized that the surfaces of various bodies absorb some portions of white light and reflect other parts.

This is why we see the world in colour: various materials reflect different parts of the spectrum, and our senses generate the feeling of seeing colourful objects, while the objects themselves have no colour of their own. Newton repeatedly writes that he does not understand that phenomenon, but he knew that the colours of bodies are only our "projection," not that body's nature. (On the other hand, he did recognize some real properties of bodies, such as their spatial extent, movement, impenetrability, inertia and weight.)[205] He also believed that mass had an objective microstructure. Such a *phenomenalistic* approach is typical for Newton and will play a role in our analysis.

About mathematics: The invention of differential *calculus* is shared by Newton and Leibniz. Both scientists waged a long argument for its priority.[206]

Nowadays, it seems that Newton approached mathematics by studying tangents to curves – i.e. he first used *derivations*, while Leibniz first calculated areas; in other words, he primarily *integrated*. We also know that both inventors used the work of the same predecessors.

---

204 "Newton's attempts to mathematize phenomena like diffraction, birefringence, and periodicity failed and few would doubt that the *Opticks* was very much a book of its time and nation, while the *Principia* is a work of timeless immortality belonging to all of numerate mankind." A. Ruppert Hall: *All was light: An Introduction to Newton's* Opticks, Oxford: Clarendon Press of Oxford University Press, 1993, p. 32.

205 Wolfgang Röd: *Novověká filosofie II*, Praha: Oikoymenh, 2004, p. 19.

206 That time sequence has been thoroughly studied at the Kyoto University.

On this topic, consult Ivan Saxl's paper[207] on medieval mathematical physics and its founders from the 13th century. Some inspiration for Newton's differential calculus certainly may have come from Maimonides' "Natura Non Facit Saltus": nature's gradual infinitely small steps.

De Gravitatione

Newton's groundbreaking work *Philosophiae Naturalis Principia Mathematica*, known as the *Principia* for short, was first published in 1687. Newton was encouraged by Edmund Halley who recognized that it would be a revolutionary piece of work.[208]

Shortly before writing his *Principia,* in year 1684 Newton wrote *De Gravitatione et æquipondio fluidorum.*[209] It remained in manuscript form, but it showed his way of thinking. We present here a few ideas from this text. They illustrate how Newton progressed toward the perfection of *Principia*. In the *De Gravitatione*, the difference between scientific and speculative way of thinking is not yet obvious. Furthermore, we may follow the gradual differences of opinion between Newton and Descartes: "Newton may not have broken decisively with Cartesianism until the eve of composing his own *Principia*."[210] We now see that this departure was crucial for the progress of Newton's method.

McGuire does not hesitate to call Newton's work from that period *physicotheology*,[211] but perhaps an even better name would be *theophysics*.

God and His function in this world are, according to Newton, most significant. Let us remember that while he worked on both his *De Gravitatione* and his *Principia*, Newton still carried out intensive alchemical experiments; therefore alchemical inspirations are quite probable in those papers.

---

207 Ivan Saxl: *Thomas Bradwardine and other Oxford's mathematicians*, manuscript in Czech, 2006.

208 There was some danger that Robert Hook, too, might claim authorship of some of Newton's results. Since Newton already had such conflict with Leibniz, he overcame his hesitation and published his work against his concern for possible negative reception. His fears eventually proved justified.

209 MS. 4003, Cambridge University Library. Available at Newtonproject: http://www.newtonproject.sussex.ac.uk/texts/viewtext.php?id=THEM00093&mode=normalized.

210 James E. McGuire: "The Fate of the Date: The Theology of Newton's *Principia* Revisited," in: *Rethinking the Scientific Revolution*, Margaret J. Osler (ed.), Cambridge: Cambridge University Press 2000, pp. 271–296, p. 279. Newton obviously did not part from Descartes until the eve of writing his *Pricipia*.

211 Ibid., p. 278.

In the manuscript of *De Gravitatione* Newton first tells us what he will not define, since it is widely known. Then, after some initial succinct definitions, he gives his concept of body, force and gravity, space and geometry.[212] We present them briefly in this order.

Definitions
Terms like *quantity*, *duration* and *space* are well known and therefore need not any other definition.
Def. 1: Place is a part of space which is eventually filled with something
Def. 2: Body is this, which fills the space.
Def. 3: Rest means to stay at the same place.
Def. 4: Motion means the change of place.[213]

Newton first parted from Descartes over the definition of a "body": Descartes took the body's extent as fundamental:

It also seems that Descartes, in Part II, paragraph 4 and 11, demonstrated that a body cannot be distinguished from its dimensions, while we may abstract from its hardness, colour, weight, temperature and other qualities that the body does not have to possess; thus what remains is its extent in length, width and depth.[214]

Newton disagreed with Descartes. He wrote in his reflections:

It remains to specify the nature of the body. This explanation, however, be less accurate because there is no necessity, but from God's will. [...] It is certain that God can stimulate our perception of their own will, and therefore may also exercise this power on the results of his will. [...] We can define the body as a certain amount of extensionality that omnipresent God endows with certain conditions:

212 We quote here the pivotal sentences from *De Gravitatione* and we also quote the conclusion made by McGuire.

213 *De Gravitatione*, Definitiones
Nomina quantitatis, durationis et spatij notiora sunt quam ut per alias voces definiri possint.
Def. 1: Locus est spatij pars quam res adæquate implet.
Def. 2: Corpus est id quod locum implet.
Def. 3: Quies est in eodem loco permansio.
Def. 4: Motus est loci mutatio

214 *De Gravitatione*: Cæterum cum Cartesius in Art 4 & 11 Part 2 Princip demonstrasse videtur quod corpus nil differt ab extensione; abstrahendo scilicet duritiem, colorem, gravitatem, frigus, calorem cæterasque qualitates quibus corpus carere possit ut tandem unica maneat ejus extensio in longum latum et profundum quæ proinde sola ad essentiam ejus pertinebit.

1) That they could move [...]
2) That two cases of this kind (i.e. two bodies) can not occur simultaneously on the same place [...]
3) That they may excite different kinds of perception and imagination in the minds of created beings.[215]

This needs a summary and comment: a body owes its existence to God who provides it with necessary space. The body's dimension is not its essential property, (as Descartes says), but only a part of space occupied by that body.

McGuire wrote: "we can define bodies as determined quantities of extension which omnipresent God endows with certain qualities."[216]

This may be the crucial point for the progress of modern science. Here is why we may replace a spatial body by a mere *point*. Newton stressed and repeated several times in his *De Gravitatione* that dimensionality is not important for a body as Descartes thought; what is important are the body's properties; mainly its capability of exerting forces upon other bodies. Elsewhere, Newton emphasizes that space and its parts do not move.

Parts of space are immovable. – Partes spatij sunt immobiles.[217]

If Newton wanted to study the movement of bodies, he could not define them by the space which they occupied; but he could (perhaps had to) replace a body by a "particle," a mass point. Here we quote Henry More who may have contributed to this school of thought by his emphasis on God's omnipresence:

How indeed could He (God) communicate motion to matter, [...] if He did not touch the matter of the universe in practically the closest manner, or at least had not touched it at a certain time? Which certainly He would never

215 Ibid. [N]atura corporea ex altera parte restat explicanda. Hujus autem, cùm non necessario sed voluntate divina existit [...] Nam certum est Deum voluntate suâ posse nostras perceptiones movere, et proinde talem potestatem effectibus suæ voluntatis adnectere [...] corpora definire possemus esse Extensionis quantitates determinatas quas Dius ubique præsens conditionibus quibusdam aificit:
1) Ut sint mobiles [...]
2) Ut ejusmodi duo non possint qualibet ex parte coincidere [...]
3) Ut in mentibus creatis possint excitare varias sensuum et phantasiæ perceptions.

216 James E. McGuire: The Fate of the Date, pp. 280–1.

217 *De Gravitatione.*

be able to do if He were not present everywhere and did not occupy all the spaces? [...] I believe it to be clear that God is extended in His manner just because He is omnipresent and occupies intimately the whole machine of the world, as well as its singular particles.[218]

Therefore God's omnipresence permits a meaningful substitution of a body by a point with the same force manifestations.

Force is then in *De Gravitatione* defined as follows:

Def. 5: Force is causal beginning for a move and a rest. It is both external, which creates or destroys or otherwise changes the caused movement of the body, or is it an internal principle, which the existing peace or movement in the body is maintained with. [...][219]

The manuscript then defines several kinds of forces: No. 6, Conation; No. 7, Impact; No. 8, Inertia; No. 9, Pressure; and, finally, No. 10, Gravity.

Def. 10: Gravity is the force in the body, which forces it to fall. Here, however, the fallen is not only meant the movement towards the centre of the Earth, but also toward any centre or any region or even from any point.[220]

This definition shows that gravity, in Newton's thought, is leaving the only domain of Earth and is becoming more general force. Regarding the concept of space and geometry, again according to Newton, there is a close ontological link to God. McGuire summarizes the characteristics of space according to Newton as follows:

(1) Space exists because God exists and is coeternal with God omnipresent being; it is neither a substance nor an accident but, in combination with absolute time, space possessed a mode of being that is presupposed by all other forms of being;

218 John Henry: A Cambridge Platonist's materialism: Henry More. *Journal of the Warburg and Courtald Institutes*, xlix (1986), p. 266.

219 *De Gravitatione*. Def. 5: Vis est motus et quietis causale principium. Estque vel externum quod in aliquod corpus impressum motum ejus vel generat vel destruit, vel aliquo saltem modo mutat; vel est internum principium quo motus vel quies corpori indita conservatur.

220 Ibid. No. 6, Conatus; No. 7, Impetus; No. 8, Inertia; No. 9, Pressio; and, finally, No. 10, Gravitas.
Def. 10: Gravitas est vis corpori indita ad descendum incitans. Hic autem per descensum non tantum intellige motum versus centrum terræ sed et versus aliud quodvis punctum plagamve, aut etiam a puncto aliquo peractum.

(2) space is eternal and uncreated and is the receptacle, within which all created things come into being;
(3) lastly, infinite space possesses an inherent structure of geometrical solids, limitless in number and differing in size, together constituting all variety of shapes [...][221]

And McGuire summarizes Newton's concept of geometry:

In short: Newton's geometry has a definite content: it is about the *motions of physical objects* generated by real forces in absolute space and time. Physical motions therefore constitute more than a possible interpretation of geometry: they are its postulates as *constructed by God*. If we know geometry by constructing it, it is because we made it: in contrast, the "natural" geometry of the cosmos is made by God.[222]

The above definitions fully justify the use of the term *theophysics*: the manuscript of *De Gravitatione* always depends on God and his permanent creative role in the functioning world. Newton describes the origin of the cosmos as a hierophany, a sacred cosmology or a cosmology of sacred places. By divine influence, space was given an inherent structure of an infinite number of geometric shapes; those are then fulfilled by God with different clusters of properties, and thus bodies are created. Gravity is then the force which makes the structure of cosmos visible.

So much about the manuscript of *De Gravitatione*.

Principia – Newton's mature methodological approaches

In his *Principia,* (first published in 1687), Newton summarized his previous ideas and elaborated on them. Compared with *De Gravitatione*, he substantially improved their precision. This precision was one of the deciding factors in Newton's progress and success. He introduced the concept of the "mass point." Indeed, the *Principia* start with the definition of mass.

Definitio I. Quantitas materiae est mensura ejusdem orta ex illius Densitate & Magnitudine conjunctim.[223]

221 James E. McGuire: *The Fate of the Date*, pp. 279–80.

222 Ibid., p. 290.

223 Definition I. Quantity of matter is a measure of the same, arising from its density and bulk

The *Principia* are organized in *modo geometrico:* theorems are derived from definitions and axioms according to Euclid. They are even proved geometrically. Although Newton himself developed infinitesimal calculus, he did not use it in his physical proofs. That was done by Newton's followers.

Newton provided a mathematical model for real space (developed *analytic geometry*) and time (introduced *linear time*). That enabled him to move between his idealized physical system and the mathematical map of the cosmos.[224] This was the decisive step for the birth of modern science. It is a subject for further examination, of which we have presented only an outline, because it could lead to endless depths. From his work on optics, Newton was aware that the visible world is just a "projection," a creation of our mind. But he ventured describing one product of the mind by another, hoping for the best. He thus built one projection from another.

Newton introduced the concept of absolute space and time. That was, in a sense, contrary to his phenomenalism, since those entities cannot be empirically established. Of course, Newton needed those concepts so he could formulate his three Laws of motion, his *axiomata sive leges motus.*[225] Only in the appendix to the second edition of his *Principia* did he spell out exactly what that absolute space and time mean. We shall return to that topic in Chapter VI.

Newton knew that this metaphysical image transcended the strict scientific theory. That is why he did not incorporate it in the main text, but included it at the end of *Scholium Generale.*

---

conjointly. Isaac Newton: *The Principia: Mathematical principles of natural philosophy and his system of the world*, translated into English by Andrew Motte in 1729, revisited by Florian Cajori, Berkeley: University of California Press, 1974, p. 1 (p. 29 in the PDF file).

224 James E. McGuire: *The Fate of the Date*, p. 294.

225 Lex I Corpus omne perseverare in statu suo quiescendi vel movendi uniformiter in directum, nisi quatenus illud a viribus impressis cogitur statum suum mutare.

Lex II Mutationem motus proportionalem esse vi motrici impressæ, & fieri secundum lineam rectam qua vis illa imprimitur.

Lex III Actioni contrariam semper & æqualem esse reactionem: sive corporum duorum actiones in se mutuo semper esse æquales & in partes contrarias dirigi. Isaac Newton: *Philosophiae naturalis principia mathematica*, Cantabrigiae: 1713, pp. 12–13.

I. Every body continues in its state of rest, or in uniform motion in a right line, unless it is compelled to change that state by forces impelled upon it.

II. The change of motion is proportional to the motive force impressed; and is made in the direction of the right line in which that force is impressed.

III. To every action there is always opposed an equal reaction: or, the mutual actions of two bodies upon each other are always equal, and directed to contrary parts. Isaac Newton: *The Principia*, translated into English by Andrew Motte in 1729, p. 13 (p. 41 in the PDF file).

Elsewhere, he says: "Whereas the main business of Natural Philosophy is to argue from Phaenomena without feigning Hypothese and to deduce Cause from Effects [...]"[226] Newton's rejection of hypotheses must have been caused by his conflict with the Cartesians and scholastics, who distorted phenomena to suit their hypotheses. And here, by introducing the absolute space and time, Newton contradicted himself. It had to wait for Kant who cleared the issue by calling space and time concepts of the *a priori* knowledge.

Newton says explicitly that he is not looking for causes of natural phenomena: that is his famous *Hypotheses non fingo* (I do not frame hypotheses) he tried to separate scientific thinking from the metaphysical one. However, that famous sentence must not be taken out of context. Newton himself, both in the *Queries* of his *Opticks* and in the appendix to his *Principia*, in the *Classical Scholia*[227] and, eventually, in his *Scholium Generale*, speculates about the true cause of gravity, about a divine force that holds the world together.

According to Newton, the planetary system and the entire cosmos have to be constantly maintained by an intelligent force; God not only created this world, He also maintains it. So the principal idea of Newton's physics is that of an ordered universe created and maintained by God.[228]

Naturally, Newton's concept of gravity was greeted with an avalanche of objections. Newton was criticized for inventing an occult concept. And, indeed, his alchemical knowledge[229] may have helped him with the idea of *magnesia*, a universal sympathy of all things.[230]

The idea of a "unified world theory" is another metaphysical concept introduced by Newton. It is an automatic assumption of modern science, yet it is not at all self-evident.

Here we just mention one of the conclusions of Chapter III: a strict monotheism is the only religion that permits the concept of a unified functioning of the world. A return to polytheism, in which Newton also included the Christian doctrine of Trinity, would make scientific research impossible, because the unity of the world would be lost. As we said in the chapter about Newton the theologian, he studied two books all

226 Query XXVIII, Optics 1718.

227 McGuire, J. E. and Rattansi, P. M.: "Newton and the 'Pipes of Pan,'" in: *Notes and Records of the Royal Society of London*, Vol. 21, No. 2 (Dec. 1966), pp. 108–142.

228 Wolfgang Röd: *Novověká filosofie II*, Praha: Oikoymenh, 2004, p. 29.

229 See the above mentioned book by B. J. Teeter Dobbs.

230 Another name for gravity is *Harmonia Mundi*, or sympathy of all things, i.e. a love in the *cosmic sense*.

his life: the book of God's word, and the book of God's work, and he searched for truth in the unity of the Bible and the created world. Newton's lifelong intention was to search for God. He tried to decipher God's plan everywhere. And Nature offered the material for his study: "There is no way (without revelation) to come to the knowledge of a Deity but by the frame of nature." [231]

The principal concept of Newton's physics is his phenomenalistic approach to the world. Newton did not invent that approach; but that step is so important that we shall return to it in Chapter VI and elaborate on it.

The unity of the world had another consequence which was absent in Galileo and others: Galileo's exploration of natural phenomena was separate from ethics and faith. But for Newton, nature with its phenomena points toward God: therefore, toward us, nature exerts a moral appeal:

> If natural Philosophy in all its Parts, by pursuing this Method [i.e., experiment], shall at length be perfected, the Bounds of Moral Philosophy will be also enlarged. For so far as we can know by natural Philosophy what is the first Cause, what Power he has over us, and what Benefits we receive from him, so far our Duty towards him, as well as that towards one another, will appear to us by the Light of Nature.[232]

Newton used Paracelsus' term *Lumen Naturae*, Light of Nature verbatim. If he was even here inspired by the mystical Paracelsus, then it is one more piece of evidence that his world was not cool clockwork, but a living nature, with purpose and moral responsibility.

Whenever Newton mentions moral appeal, we have to expect the influence of alchemy and theology.

We have already mentioned that Newton's achievements were due to his methodological superiority over his contemporaries. It was based on mutual balancing and correcting the various schools of thought. So, in conclusion, let us quote Newton's recommendations for research in natural philosophy in his *Regulae philosophandi* at the beginning of book III in the Second Edition (1713) of his *Principia*:

> Regula I. No more causes of natural things should be admitted than are both true and sufficient to explain their phenomena.

231 Newton, Yahuda MS 41, f. 7r.

232 Items 1238–1243, acc. to book by John Harrison: *The library of Isaac Newton*, pp. 209–10.

> Regula II. Therefore, this causes assigned effects of the same kind must be, so far as possible, the same.
> Regula III. Those qualities of bodies that cannot be intended (i.e. qualities that cannot be increased and diminished) and that belong to all bodies on which experiments can be made, are to be considered the universal qualities of all bodies whatsoever.[233]

Rules No. I and II spell out the principles of parsimony and simplicity. Newton's Nature is simple; it does not do anything superfluous. Rule III has an ontologic meaning: it leads us to the distinction between *things as they are* and our sensory perceptions of those things. Strictly qualitative (intensive) properties must not be ascribed to things themselves. That may be done only to their extensive properties, this is the spot where Newton's "induction" appears: he depended on it and freely used it in his Rules. For instance, G. A. J. Rogers wrote: "The key to Newton's strong faith in the success of inductive experimental practice was the same "God of order" who has so structured nature that the experimentalist can assume simplicity. God Himself guarantees induction."[234]

Therefore, put simply, we may depend on God's help: by means of an inductive reasoning, He enables us to find truth in this world.

Of course, this creates a logical conflict: on the one hand, Newton believed in a *phenomenological approach* and in *induction*; but, on the other hand, he needed a *metaphysical* framework for his scientific theories (later Kant's *a priori* knowledge). "[W]hereas the main business of Natural Philosophy is to argue from Phaenomena without feigning Hypothese and to deduce Cause from Effects [...]"[235]

But recognizing this framework means giving up an induction. Röd explicitly writes: "As soon as Newton's physics rests on a metaphysical framework that cannot be derived from experience, Newton's program is no longer in harmony with his physics."[236] All this combining and balan-

233 Isaac Newton: *Philosophiae Naturalis Principia Mathematica*, London: 1713, p. 357.
Regula I. Causas rerum naturalium non plures debere, quam quæ & vere sint & earum Phænomenis explicandis sufficiant.
Regula II. Ideoque Effectuum naturalium ejusdem generis eadem sunt causæ, fieri potest.
Regula III. Qualitates corporum quæintendi & remitti nequement, quæque corporibus omnibus competunt in quibus experimenta instituere licet, pro qualitatibus corporum universorum habendæ sunt.

234 G. A. J. Rogers: "Newton and the guaranteeing God," in Force and Popkin (eds.): *Newton and Religion*, Dordercht: Kluwer Academic Publishers 1999, pp. 221–35.

235 Query XXVIII, *Opticks* 1718.

236 Wolfgang Röd: *Novověká filosofie II*, p. 30.

cing of various approaches seems to have left behind more visible "seams".

On the one hand, Newton was a phenomenologist; on the other hand, he crossed that line when he needed metaphysical speculations as boundary conditions for his scientific theories.

Nevertheless, Newton's physics, as a solidly based quantitative science, offered the foundations for a description of phenomena and was also able to accurately predict them. It was successful in its objective. Newton's laws of motion and theory of gravitation explained not only the orbits of planets, but also the tides, motion of the Moon under the influence of the Sun and Earth, orbits of comets, perturbations in the orbits of Jupiter and Saturn, even the precession of the equinox.

Summary of Chapter V

We shall try to list Newton's principal ideas from which he created his physics. Parenthetically, we mention the extra-physical disciplines which, in our opinion, gave him the inspiration.

1) This world functions as a single unit because it is based on the concept of a single God. (Theology.)
2) Newton introduces absolute space and time that somehow depend on God. That connection is clearly spelled out in *Scholium Generale.* (See Chapter VI.)
   The linear time comes from the postulate of historicity and follows from the conviction that there is an eternal One, transcending time and being revealed through history. At this point, he is clearly inspired by Maimonides. (Theology and history.)
   Absolute space contains an infinite number of inherent geometrical forms and trajectories, prepared by God. Thus the real (absolute) space may be described by the means of analytical geometry.
3) A body arises when God grants a set of properties to some portion of space. A body is then God's work: it has two main groups of properties that it applies toward its environment: force and the ability to provoke (excite?) our sensory perceptions. (Theology – God is the creator.)
4) A phenomenalistic approach, based on our perceptions, represents our examination of God's action. (To be discussed in Chapter VI.)
   This approach allows us to separate the world as such, which is a sacred work of God, from the veil of phenomena that may be studied factually, rationally and mathematically.

5) Given that the size of a body is not essential, it is permissible to replace the body by a mass point with the force effects of the entire body. That opens the way to switching there and back between mathematically described space and physical models. It is a fundamental novelty of Newton's physics.
6) Gravity makes the inherent structure of the cosmos visible. It expresses the coherence of the world as a whole; it is analogous to the convergence of the transcendent pole toward God. God is the centre of centres (Nicolaus Cusanus). Gravity may be interpreted as God's omnipresence, since all bodies attract each other. That is an old hermetic idea. In theological terms, it is Love. In alchemistic terms, it is *magnesia*. (Theology and alchemy.)
7) The world can be described by natural laws, because it is designed and powered by an intelligent being. Here Newton could feel as a new Moses, through whom God reveals new laws of nature.
8) The application of quantification and mathematics is justified for Newton by the apocryphal Biblical Book of Wisdom 11,20: "...but thou hast ordered all things in measure, and number and weight." (Theology.)
9) Our ability to comprehend the world, as well as the justification of the induction method, is guaranteed by God's grace that enables us to truly know his work. (Theology.)

We conclude this section by stating that the terms *physicotheology* or *theophysics* are justified. Our further study – an analysis of the *Scholium Generale* – will therefore focus on obtaining the most detailed image of Newton's God.

VI. Analysis of the *Scholium Generale*

Phenomenal character of the world and all being, if it is the subject of science, however, arises from the knowledge that is in fact not of scientific nature.
Karl Jaspers[237]

Newton's pivotal work, *Philosophiae Naturalis Principia Mathematica*, was written in Latin and first published in 1687. For its second edition of 1713, apart from some minor corrections, Newton added a chapter that was to put the entire work into a broader intellectual context. He wrote two versions of that supplement. First one was an addition to the Third Book of the *Principia*, *Proposition IV* through *IX*. It was not used and remained in manuscript. Today, this text is referred to as *Classical Scholia*, because of its many references to authors of antiquity. Newton tried to show that the basic ideas of his physical and mathematical studies were already known to the ancient authors.

Classical Scholia has been the subject of a thorough examination by a number of authors. Best known among the analyses are probably "Newton and the 'Pipes of Pan'" by McGuire and Rattansi[238] and "Newton: The *Classical Scholia*" by Paolo Casini.[239] *Classical Scholia* is a very extensive text. In connection with our focus on *Scholium Generale*, let us just mention that it is obviously strongly inspired by Ralph Cudworth. *Classical Scholia* may even seem to be a parallel to Cudworth's *The True Intellectual System of the Universe*: Cudworth wanted to prove that all the major nations of antiquity were, in their own way, monotheistic, that

237 Karl Jaspers: *Chiffren der Transcendenz* (in Czech *Šifry transcendence*, Praha: Vyšehrad 2000, p. 40.)
238 J. E. McGuire, P. M. Rattansi: "Newton and the 'Pipes of Pan'", in: *Notes and Records of the Royal Society of London*, Vol. 21, No. 2 (Dec. 1966), pp. 108–143.
239 Casini, Paolo: "Newton: The Classical Scholia," in: *History of Science*, XXII (1984), pp. 1–58.

they all recognized one supreme God. Likewise, Newton tried to prove that ancient scholars had known about heliocentrism, gravity, internal structure of matter etc. And, like Cudworth, he did so by quoting a large number of ancient authors.

However, finally, Newton used the *Scholium* as the conclusion of his *Principia* quite differently. He felt that he had no need for ancient philosophers. *Scholium Generale* (further *S.G.*) has become an "abstract" of the *Principia*; it summarizes the natural-philosophical, methodological, metaphysical and religious opinions of their author. In the *Principia*, those opinions are included implicitly.[240]

At the time of the second edition of the *Principia*, Newton was already seventy years old, he had attained a significant social status, and was even ennobled. He was still intellectually active, though no longer in the field of natural philosophy. He turned his interest to areas which we would today probably call economics. He became the Master of the Royal Mint. And he was fully occupied by his historical and theological studies, too. So the work of his last few years is perhaps his spiritual testament. It exposes his ideas at the end of his life, reveals his deepest intentions, and reveals the way of his thinking in the times of his greatest mental powers.

Scholium Generale is a peculiar text. It is extremely dense and complex. It consists of just five paragraphs, of altogether 1450 words. Newton managed to discuss a multitude of topics there, including gravity, planetary systems, the movement of comets, space and time, but also his *design arguments* for the *creationess* of the world, the tides, active power, electricity, even the nervous system. He refuted Cartesianism with its hypothesis of vortices,[241] swept the views of Leibniz, justified the true purpose of natural philosophy, and defended his own methodology based on induction. However, the text includes much more. It is not just one of the most complex, most famous and most polemical texts in the history of science; its main themes are Newton's ideas about God. Those

240 However, Bernard Cohen has shown that a natural theology was present right from the start in all editions of *Principia*, not only implicitely, but explicitely. Thus, in Corollatium 4 of Proposition VIII in Book III, Newton writes: "God therefore placed the planets at different distances from the Sun so that according to their degrees of density they may enjoy a greater or less proportion of the Sun's heat." See Bernard Cohen: Isaac Newton's *Principia*, the Scriptures, and the divine providence, in *Philosophy, science, and method*, ed. Sidney Morgenbesser, et al. (New York: St. Martin's Press, 1969), p. 523–48.

241 In Chapter V. we have already mentioned his departure from Descartes in connection with the manuscript *De Gravitatione*. This also shows that the *Scholium Generale* is not just a subsequent justification of the *Principia*, written more than twenty years after their first edition. Many ideas of the *Scholium Generale* can be found in the manuscript from the year 1684.

ideas go far beyond the scope of similar texts of other contemporary authors. We believe that it is Newton's most concise and most honest text on that subject. It reveals his true beliefs about God and the world. And it most conclusively shows the pattern of thought from which Newton's science grew.

S.G. is a well-known text. However, to what extent it has been comprehended, is questionable. Even its most famous phrase, "hypotheses non fingo," presents ambiguity. If it is used out of context, it can substantially distort the image of its author.

Many ideas of the *S.G.* may be found in a more comprehensive form in Newton's other private manuscripts, not intended for publication. They are now a significant help in the interpretation of *S.G.*, which is sometimes so dense that it borders on obscurity.

Today it is widely agreed that the principal purpose of the *S.G.* can be expressed as follows: the laws of physics reveal that the universe must have been built according to a plan. It is absolutely impossible that such a perfect object would have been developed merely by blind chance (*the design argument*). And so Newton wrote his *Principia* with the intention of making the creation of the world more comprehensible. His natural philosophy was to become a kind of *corollary* to the Biblical texts. Essentially, he created an apologetic work which brought a rational that intention, therefore undeniable, proof of the existence of God. In the *S.G.* is expressed several times, always very brief and concise. For example:

> Elegantissima hacce solis, planetarum & cometarum compages non nisi consilio & dominio entis intelligentis & potentis oriri potuit.[242]

To understand such a profound statement more thoroughly, we present a few more of Newton's remarks on the subject. For example, in his now well-known and often quoted letter of 10 December 1692,[243] addressed to Richard Bentley, Newton is more explicit about his ideas:

242 Isaac Newton: *Philosophiae naturalis principia mathematica*, Cantabrigiae MDCCXIII, p. 482. This most beautiful System of the Sun, Planets, and Comets, could only proceed from the counsel and dominion of an intelligent and powerful being.
English translation: *The mathematical principles of natural philosophy by Sir Isaac Newton*: translated into English by Andrew Motte, 1729, 2 vols. reprinted with an introduction by I. Bernard Cohen, London: Dawsons, 1968, vol. 2, p. 388.

243 That letter dates before the year 1693 when Newton suffered a nervous breakdown, which some scholars would like to see as the cause of his sudden turn toward theology. However, that argument is faulty.

Sir
When I wrote my treatise about our Systeme I had an eye upon such Principles as might work with considering men for the beliefe of a Deity & nothing can rejoyce me more then to find it usefull for that purpose But if I have done the publick any service this way 'tis due to nothing but industry & a patient thought. [...]
...if the Sun was at first an opake body like the Planets or the Planets lucid bodies like the Sun, how he alone should be changed into a shining body whilst all they continue opake or all they be changed into opake ones whilst he remains unchanged, I do not think explicable by mere natural causes but am forced to ascribe it to the counsel & contrivance of a voluntary Agent.
The same power whether natural or supernatural, which placed the Sun in the center of the orbs of the six primary Planets, placed Saturn in the center of the orbs of his five secondary Planets & Iupiter in the center of the orbs of his four secondary ones & the earth in the center of the Moons orb; & therefore had this cause been a blind one without contrivance & designe the Sun would have been a body of the same kind with Saturn Iupiter & the earth, that is without light & heat. Why there is one body in our Systeme qualified to give light & heat to all the rest I know no reason but because the author of the Systeme thought it convenient, & why there is but one body of this kind I know no reason but because one was sufficient to warm & enlighten all the rest. [...]
To your second Query I answer that the motions which the Planets now have could not spring from any naturall cause alone but were imprest by an intelligent Agent. [...]
This must have been the effect of Counsel. Nor is there any natural cause which could give the Planets those just degrees of velocity in proportion to their distances from the Sun & other central bodies about which they move & to the quantity of matter conteined in those bodies, which were requisite to make them move in concentrick orbs about those bodies. Had the Planets been as swift as Comets in proportion to their distances from the Sun (as they would have been, had their motions been caused by their gravity, whereby the matter at the first formation of the Planets might fall from the remotest regions towards the Sun) they would not move in concentric orbs but in such excentric ones as the Comets move in. Were all the Planets as swift as Mercury or as slow as Saturn or his Satellites, or were their several velocities otherwise much greater or less then they are (as they might have been had they arose from any other cause then their gravity) or had their distances from the centers about which they move been greater or less then they are with the same velocities; or had the quantity of matter in the Sun or in Saturn Iupiter & the

> earth & by consequence their gravitating power been greater or less then it is: the primary Planets could not have revolved about the Sun nor the secondary ones about Saturn Iupiter & the earth in concentrick circles as they do, but would have moved in Hyperbolas or Parabolas or in Ellipses very excentric. To make this systeme therefore with all its motions, required a Cause which understood & compared together the quantities of matter in the several bodies of the Sun & Planets & the gravitating powers resulting from thence, the several distances of the primary Planets from the Sun & secondary ones from Saturn Iupiter & the earth, & the velocities with which these Planets could revolve at those distances about those quantities of matter in the central bodies. And to compare & adjust all these things together in so great a variety of bodies argues that cause to be not blind & fortuitous, but very well skilled in Mechanicks & Geometry. [...]
> There is yet another argument for a Deity which I take to be a very strong one, but till the principles on which tis grounded be better received I think it more advisable to let it sleep. I am
> Your most humble Servant to command
> Is. Newton.
> Cambridge Dec. 10th [244]

Another proof of Newton's ideas as to the *design argument* are some parts of his second major work, *Opticks*, especially its final part, *Queries*. It somehow resembles the *S.G.* in the form of questions. Some of those questions stretch over ten pages. The *Query* number 31 tells us, inter alia:

> Now by the help of these Principles, all material Things seem to have been composed of the hard and solid Particles above-mention'd, variously associated in the first Creation by the Counsel of an intelligent Agent. For it became him who created them to set them in order. And if he did so, it's unphilosophical to seek for any other Origin of the World, or to pretend that it might arise out of a Chaos by the mere Laws of Nature; though being once form'd, it may continue by those Laws for many Agens... Such a wonderful Uniformity in the Planetary System must be allowed the Effect of Choice.[245]

Another argument for the logical necessity of God's design, based on the consideration of symmetry, is also known from a private manuscript:

244 See http://www.newtonproject.sussex.ac.uk/view/texts/normalized/THEM00254.

245 Newton, *Opticks, or a treatise of the reflections, refractions, inflections & colours of light*, 4th ed., 1730 New York: Dover, 1952, p. 402.

> Can it be by accident that all birds beasts & men have their right side & left side alike shaped (except in their bowells) & just two eyes & no more in either side the face & just two ears on either side the [sic] head & a nose with two holes & no more between the eyes & one mouth under the nose & either two fore leggs or two wings or two arms on the sholders & two leggs on the hipps one on either side & no more? Whence arises this uniformity in all their outward shapes but from the counsel & contrivance of an Author?[246]

Shortly after the first edition of the *Principia*, Newton's contemporaries realized that Newton's scientific writings contained an abundance of data, demonstrating the need for an intelligent design for the world (*design argument*); and, therefore, that they possess a powerful apologetic potential at a time when scientific progress threatened with a complete emancipation of science from religion. It was mainly the aforementioned Richard Bentley and William Whiston, Newton's successor as Lucasian Professor of Mathematics at Cambridge. Bentley used the Arguments from the *Principia* in preparation of Boyle's lectures for publication. And Whiston was inspired to write his *Astronomical principles of religion, natural and reveal'd.* For a preface, he copied Newton's *Classical Scholia* almost verbatim. He did not give Newton credit; however, he was thus instrumental that the content of the *Scholia* was finally published. These two examples show how such proofs of God's existence and God's design of this world became fashionable in England of the 17th and 18th centuriy.

Previous Research on the *Scholium Generale*

The supremely rational text of the *Principia* ends with an explanatory chapter that in fact exceeds the powers of rational thought. There are several ways of grasping the *S.G.* We shall mention two different interpretations, and then propose one of our own.

First, following Steven D. Snobelen, one of the prominent researchers in the field of Newton's theology, *S.G.* may be understood as an *Anti-Trinitarian Manifesto*. Snobelen has published several papers along those lines.[247] He stressed that it was precisely Newton's heterodox concept

246 Newton, Keynes MS 7, p. 1.

247 Stephen D. Snobelen: *The Theology of Isaac Newton's General scholium to the Principia*, see http://www.isaac-newton.org/.
Stephen D. Snobelen: *To discourse of God: Isaac Newton's heterodox theology and his natural philosophy*, see http://www.isaac-newton.org/.

of God that made modern science possible. As a matter of fact, it even supplied the necessary motivation.

Second interpretation, proposed by Rudolf de Smet and Karin Verelst,[248] is to seek inspiration in *platonic stoicism*.

In summary, those two interpretations have been elaborated: the first one as an *anti-trinitarian* theology of the author, the second one as a *platonic stoicism*, thus a *syncretic* philosophy coming from Philo of Alexandria (perhaps via his followers, such as Justus Lipsius and Henry More.)

Our own research will build upon both of the theories mentioned above. It will confirm the anti-trinitarian position of the author, but with the caveat that it is a long way from fully exhausting Newton's heterodoxy. Likewise, it will recognize the Alexandrian *syncretic* philosophy, however – while remaining on Alexandrian soil – it will jump over to the hermetic writings of the second century AD.

Scholium Generale and *Prisca Sapientia* (Hermetic writings, Cudworth and Patrizi)

We believe that we should again stress the role of Newton's contemporaries, the Cambridge Platonists More and Cudworth, and Newton's reliance on *Prisca Sapientia*: the idea that all the wisdom of antiquity derived from Noah's original revelation. For example, Dobbs in her book *The Foundations of Newton's Alchemy* extensively investigated the Cambridge Platonists, mainly Henry More, and showed that, through them, Newton gained many ancient philosophical concepts.[249] There is no doubt that Newton studied Cudworth: we have already mentioned a short but very important manuscript, *Out of Cudworth*, showing that Newton had carefully read Cudworth's *The True Intellectual System of the Universe*[250] and made detailed excerpts from it.[251]

Stephen D. Snobelen: *Isaac Newton, heretic*, see http://www.isaac-newton.org/.

Stephen D. Snobelen: *Isaac Newton and Socianism: associations with a greater heresy*, see http://www.isaac-newton.org/.

248 Rudolf De Smet, Karin Verelst: "Newton's *Scholium Generale*: The Platonic and Stoic legacy – Philo, Ustus Lipsius, and the Cambridge Platonists," in: *History of Science*, XXXIX (2001), pp. 1–30, see http://www.shpltd.co.uk/smet-verelst.pdf.

249 Betty Jo Teeter Dobbs: *The Foundations of Newton's Alchemy*, Cambridge: Cambridge University Press, 1975, pp. 102–11.

250 The entire text is available in PDF format at http://www.cimmay.us/r_cudworth.htm.

251 See http://www.newtonproject.sussex.ac.uk/catalogue/record/THEM00118. At this address is the following description of Newton's manuscript "Out of Cudworth," signature fN563Z:

Likewise, in her other book, *The Janus Faces of Genius*, Dobbs stresses the importance of old authors for a full comprehension of the *Scholium*: in Newton's opinion, the older the author, the more trustworthy he was, because he was closer to the first revelation of divine wisdom: "Newton would have in any case preferred the more ancient authority as being closer to the fount of Truth."[252]

For a long time, Newton's proof of the *oneness* of God was thought to come from his frequently emphasised Arianism. However, even in this respect, Newton's ideas passed through a long developmental period. We have shown in the Chapter on Theology that his Arian sympathies changed radically in the course of years: such a profound and free spirit as was his, could not be expected to be satisfied by accepting somebody else's idea. He always had to find his own way.

We shall concentrate on Newton's inspiration from the Hermetic literature with which he got acquainted via the writings of the above mentioned Ralph Cudworth and Franciscus Patrizi. That subject has not yet received adequate attention.

Patizi made the entire Hermetic Corpus available in Latin,[253] a language which, to Newton, was as natural as his mother tongue. It is even possible that it was due to Patrizi that Newton decided to write the *Scholium*, because Patrizi likewise concludes several Hermetic treatises with his own explanation called *scholium*. And Cudworth helps us to comprehend the *Scholium* in its very essence: both authors, Cudworth and Newton, used rational arguments against atheism.

Rudolf De Smet and Karin Verelst barely mention Cudworth's influence. They say that Philo's "Deus est omnia," which resonates through the *Scholium Generale*, has also been emphasized by Cudworth in his work on Orphic philosophy and other sources of humanistic philosophy, from antiquity up to contemporaneity:

Notes from [Ralph] Cudworth, True Intellectual System of the Universe (London, 1678) and short excerpts from Hyginus, etc. The notes from Cudworth (in order) are from pp. 13, 12, 14, 16 17, 9, 17, 23, 23, 38, 120, 121, 124, 128, 129, 129, 211, 212, 215, 216, 222, 223, 238, 248, 248, 249, 249, 250, 251, 252, , 297, 296, 297, 299, 299, 299, 300, 547, 547, 313, 311, 312, 312, 313, 313, 314, 313, 314, 315, 315, 316, 317, 319, 321, 340, 328, 330, 413, 414, 353, 354, 371, 381, 396, 397, 401, 403, 406, 407, 551, 417, 425, 426, 462, 463, 464, 465, 455, 456, 458, 388, 553, 461, 462, 462, 463, 464, 465, 483, 479–83, 465, 466, 467, 468, 529, 538, 531, 530, 538, 539, 547, 547, 548, 549, 550, 552, 593, 593.

252 Betty Jo Teeter Dobbs: *The Janus Faces of Genius*, p. 205.

253 Franciscus Patricius: *Magia Philosophica, hoc est F. Patricii summi philosophi Zoroaster & eius 320 Oracula Chaldaica. Asclepp Dialogus. & Philosophia magna Hermetis Trismegisti... Latine reddita.* Hamburgi: 1593.

> The Philonian *deus est omnia* which so strongly resounds in the *Scholium Generale* was taken up and annotated by Cudworth through references to the *Orphica*. In doing so, Cudworth proceeded in line with the humanist tradition, i.e. with extensive references to ancient as well as contemporary sources.[254]

The authors terminate their thesis by admitting that they cannot go on examining Newton's relation to Cudworth, but that it must have been far greater than what could be found in the manuscript of *Out of Cudworth*.[255] By all means: Cudworth provided Newton with inspiration not only in subject matter, but, perhaps, even to its linguistic form. For example, the third book of Newton's Principia, *De Mundi Systemate* (in English *The System of the World*) strongly resembles Cudworth's *Systema Intellectuale huius Universi* (in English *The true intellectual System of the World.*)

This is where our own research begins. We believe that those points that to De Smet and Verelst seem to come from Philo, may be found in the Hermetic Corpus. Likewise, inspiration even for the most profound ideas on which Newton built his natural science can be found in that Corpus: it is not the concept of a *real* world, but of an observed, *phenomenal* one. The following pages compare the principal idea of God and the world of Newton with that of *Corpus Hermeticum* (*C.H.*) and of *Asclepius*.

Newton knew the full text of hermetic writings from Patrizi's book. And he also was thoroughly familiar with Cudworth's work, with his quotations and with his commentaries on passages from the *C.H.*

Hermetic writings are texts that are standing just on the boundary of mythical and rational thinking. As will be shown below, Newton was intrigued by these texts for a number of reasons. Hermetic Corpus is imbued with inner piety, which Newton shared. And their God is a benevolent Creator and Maintainer of the world. He is also a somewhat predictable God-the-Creator, and that must have suited the rational Newton just right.

We shall concentrate on a detailed analysis of *S.G.*, from line 33 through line 99 in Latin. Then, we shall confront it with Cudworth's *The True intellectual system of the Universe* (in the following *T.I.S*); we shall use all 25 quotations of Asclepius and *C.H.* listed in Cudworth's book. We supplement Cudworth's work by some quotations from Patrizi's book,

---

254 Rudolf De Smet, Karin Verelst: "Newton's *Scholium Generale*: The Platonic and Stoic legacy," in *History of Science*, XXXIX (2001), pp. 1–30.

255 Ibid., p. 13.

Quod immanifestus Deus manifestissimus est[256] and *De Providentia & Fato*.[257]

We shall always start with a Latin and English quotation from *S.G.*, then give its brief explanation, then confront it with a quotation from Cudworth,[258] then call attention to similarities and/or differences between *S.G.* and *C.H.*

Finally, we shall try to conclusively demonstrate where Newton's concept of God does, or does not, have the essential features of the God of the Corpus Hermeticum, with all the consequences following from such results.

1. This most beautiful System [...] subject to the dominion of One

> **Elegantissima hacce solis, planetarum & cometarum compages non nisi consilio & dominio entis intelligentis & potentis oriri potuit. Et si stella fixa sint centra similium systematum, hac omnia simili consilio constructa suberunt Unius dominio.** This most beautiful System of the Sun, Planets, and Comets, could only proceed from the counsel and dominion of an intelligent and powerful being. And if the fixed Stars are the centers of other like systems, these, being form'd by the like wise counsel, must be all subject to the dominion of One [...][259]

Here we find the *design argument.* It was Newton's basic attitude which he mentions in many places, among them in his letter to Bentley, also in his Queries in Optics and elsewhere. For comparison, let us quote *C.H.*, Book V:

> [H]e appeareth in all and by all [...] But if thou wilt see him, consider and understand the Sun, consider the course of the Moon, consider the order of the Stars. Who is he that keepeth order? For all order is circumscribed or terminated in number and place [...] for there is some body, O Tat, that is

256 Patricius, Franciscus: *Magia Philosophica*, pp. 134–139. That the unrevealed God is the most revealed one.

257 Franciscus: *Magia Philosophica*, pp. 220–226. On Providence and Fate.

258 Wherever Cudworth quotes Asclepius, (available in Latin only), we present both Latin version and Cudworth's English translation, quotations from the Greek Corpus Hermeticum we use only Cudworth's English translation. Numbers in parentheses correspond to quotations listed in Appendix 3. Quotes from Corpus Hermeticum, beyond those used by Cudworth, come from the Czech translation by Zdeněk Kratochvíl, published online at www.fysis.cz. To facilitate reading, we follow the graphic pattern: Latin text is presented in **block** script; English text in regular.

259 The English translation of Newton's *Scholium Generale* is that of Andrew Motte (1729).

> the Maker and Lord of these things. For it is impossible, O Son, that either place, or number, or measure, should be observed without a Maker. For no order can be made by disorder or disproportion.[260]

That passage contains a parallel to the apocryphal *Book of Wisdom* 11,20, where we read: "but thou hast ordered all things in measure, and number and weight." Such words, whether form *C.H.*, or – almost verbatim – from the *Book of Wisdom*, could have been Newton's credo, his central motive for all his activities and a principal support for his *design argument:* The order and mathematical plan of this world is a proof of an Act of God.

The Greek word KOSMOS means order, rule, beauty, ornament. The apocryphal Book of Wisdom comes from the same place and, approximately, from the same period from which we have the extant *C.H.*: Alexandria about the beginning of our era.

As to the other part of the quote, i.e. the subordination to the power of a single being, we quote from Ralph Cudworth's *The True Intellectual System of the Universe* (*T.I.S.*), *C.H.* Book XII (20):

> This whole world is intimately united to him, and observing the order and will of its father, hath the fullness of life in it; and there is nothing in it through eternity which does not live; for there neither is, nor hath been, nor shall be, any thing dead in the world.
> And Cudworth adds his commentary: "The meaning is, that all things vitally depend upon the Deity."[261]

For the sake of completeness, let us include *T.I.S.*, Asclepius (1b):

> **Huius itaque, qui est unus omnia, vel ipse est Creator omnium, in tota hac disputatione curato meminisse.** Be thou therefore mindful in this whole disputation of him, who is one and all things, or was the creator of all.[262]

We then conclude the first part of our analysis recognizing that both *S.G.* and *C.H.* see the universe as an ingenious mechanism that was designed, and is now completely controlled, by the power of its Creator.

260 John Everard: *The Divine Pymander in XVII books*, London, 1650. This was translated by John Everard from the Ficino Latin translation.

261 Ralph Cudworth: *The True Intellectual System of the Universe*, p. 591.

262 Ibid., p. 588.

2. Lord God Pantokrator

> **Hic omnia regit non ut anima mundi, sed ut universorum dominus. Et propter dominium suum, dominus deus Παντοκρατωρ dici solet.** This Being governs all things, not as the soul of the world, but as Lord over all: And on account of his dominion he is want to be called Lord God Pantokrator.

Calling God *Pantocrator* is not unusual in ancient writings. Greek documents mostly use the word KYRIOS, Lord. In connection with Newton's views, there is an important theological point: God is not the soul of the Universe. God as a soul of the Universe, *anima mundi,* is the God of the pantheists, such as Giordano Bruno. It would be an immanent God identical with the world. One cannot pray to such a God. There would be no place for a life after death, no Last Judgment, etc. In Cudworth's writings, whenever we find references to *C.H.* with the designation of God as a Lord, it means that this God is *transcendent* toward the World; He has His own existence transcending this World. It is an existence with all its logical consequences: e.g., there can be only one such God. For our work, it is important that Cudworth uses terms such as the Lord for God, in many places where he deals with old Egyptian religion and in places preceding the hermetic writings. For example:

> In like manner, Horus Apollo in his Hieroglyphics tells us, that the Egyptians, acknowledging a Παντοκρατωρ *and* Κοσμοκρατορ, an omnipotent being that was the governor of the whole world, did symbolically represent him by a serpent [...][263]

Although many such quotations could be found, we include just a few. From their context, it is obvious that the God of the hermetic documents is truly a Lord, omnipotent, yet kind and beneficent. Such a Lord is worthy of respect and praise, best rendered by an apotheotic prayer: *T.I.S.*, *C.H.* Treatise XIII (22a):

> I am about to praise the Lord of the creation [...][264]

Also, *T.I.S.*, *C.H.* Treatise XVI (24):

263 Ibid., p. 566.
264 Ibid., p. 591.

I will begin with a prayer to him, who is the Lord, and maker, and father, and bound of all things [...][265]

In the last quotation, we notice that *C.H.* knows God as *father*. It is a different, deeper relationship to God as a person. For only with a person can we have a human relationship, whether it is respect, obedience, praise, or even fear. Such a relationship with a soul of the Universe is impossible.

3. God is a relative word

Nam deus est vox relativa & ad servos refertur: & deitas est dominatio dei, non in corpus proprium, uti sentiunt quibus deus est anima mundi, sed in servos. For God is a relative word, and has a respect to servants; and Deity is the dominion of God, not over his own body, as those imagine who fancy God to be the soul of the world, but over servants.

This short statement needs additional explanation of the word *relative*. It means that it is not an absolute word, therefore not God's proper name. Here, Newton touches on one of God's attributes, traditionally credited only to the Jewish God, namely that God is nameless. The Old Testament God, in Exodus 3,14, calls Himself EHEJEH ASHER AHEJEH, I am who I am. However, we also find that attribute in *C.H.*, such as *T.I.S.*, Asclepius (3):

Non spero totius majestatis effectorem, omnium rerum patrem vel dominum, uno posse quamvis e multis composito nomine nuncupari. Hunc voca potius omni nomine, siquidem sit unus et omnia; ut necesse sit aut omnia ipsius nomine, aut ipsum omnium nomine nuncupari. [...] I cannot hope sufficiently express the author of majesty, and the father and lord of all things, by any one name, though compounded of never so many names. Call him therefore by every name, forasmuch as he is one and all things; so that of necessity. Either all things must be called by his name, or he by the names of all things.[266]

And further in *T.I.S.*, *C.H.* Treatise V (8):

265 Ibid., p. 591–2.
266 Ibid., p. 588.

> He is all things that are, and therefore he hath all names, because all things are from one father and therefore he hath no name, because he is the father of all things.[267]

Therefore the supreme God must be nameless. His majesty cannot be subjected to any name. Moreover, since He is in all things, He could as well carry all names. *C.H.* Book V actually begins with this passage:

> This Discourse I will also make to thee, O Tat, that thou mayest not be ignorant of the more excellent Name of God.

The question of when God's namelessness first appeared, whether in the Bible or in the hermetic writings, and therefore which one of those sources is older, will be discussed at the end of Chapter VI, in part *Corpus Hermeticum II*. However, already Cudworth recognized from Greek sources only that the supreme God of the ancient Egyptians had various names, such as Amun or Isis, but that He was also worshipped as a nameless divine power. He quotes *Apuleius*:

> She [Isis] was an universal Deity, comprehensive of the whole nature of things, the one supreme God, worshipped by the Pagans under several names, and with different rites.[268]

Newton refers to the word *relative* in another context: the word means a *relation,* and in the case of God, it is the relation between a Lord and his servants: servants must obey and praise their lord. This point is stressed in the *Twelve Articles* more clearly then in the *S.G.*:

> Article 8. We are to return thanks to the father alone for creating us and giving us food and raiment and other blessings of this life [...]

As for praising God, we again quote *T.I.S.*, Treatise XIII (22a):

> I am about to praise the Lord of the creation.

T.I.S., Treatise XV (22b):

267 Ibid., p. 589.
268 Ibid., p. 593.

All the powers, that are in me, praise the one and the all. [269]

And another quotation from *T.I.S.*, Treatise XVI (24), mentioned above:

I will begin with a prayer to him, who is the Lord, and maker, and father, and bound of all things [...]

We thus see another correspondence between *S.G.* and *C.H.*: God is nameless, or *all-name*. The word God is not a proper name, but a description of the position of absolute superiority.

4. Supreme, eternal, infinite, absolutely perfect

Deus summus est ens aeternum, infinitum, absolute perfectum: sed ens utcunque perfectum sine dominio non est dominus deus. The supreme God is a Being eternal, infinite, absolutely perfect; but a being, however perfect, without dominion, cannot be said to be Lord God.

This is a list of superlatives defining divine qualities. We shall not dwell on them, for there is a plethora of such statements in the literature and nothing can be proven from them. We use only one quotation taken from the hermetic text.

T.I.S., Asclepius (4):

Solus deus ipse in se, et a se, et circum se, totus est plenus atque perfectus. God alone in himself, and from himself, and about himself, is altogether perfect.[270]

There is complete agreement between *S.G.* and *C.H.* Nevertheless, it does not prove anything of consequence, because God's omnipotence appears frequently and in the writings of many authors.

5. The God of *Israel*

Dicimus enim deus meus, deus vester, deus Israelis, deus deorum, & dominus dominorum: sed non dicimus aeternus meus, aeternus vester, aeternus

269 Ibid., p. 591.
270 Ibid., p. 588.

Israelis, aeternus deorum; non dicimus infinitus meus, vel perfectus meus. Ha appellationes relationem non habent ad servos. [F]or we say, my God, your God, the God of *Israel*, the God of Gods, and Lord of Lords; but we do not say, my Eternal, your Eternal, the Eternal of *Israel*, the Eternal of Gods; we do not say, my Infinite, or my Perfect: These are titles which have no respect to servants.

This shows that although Newton – according to our hypothesis – was inspired by *prisca sapientia,* he thought that his was the original concept of Noah's God of old Israel, preserved in its purest form in the hermetic documents.

This place, (like point 2 above) also shows that Newton was familiar with Cudworth and his views, and accepted them.

6. The word God signifies *Lord*

Vox deus passim significat dominum: sed omnis dominus non est deus. The word God usually signifies Lord; but every lord is not a God.

These words recount what was said in points 2 and 3. It also brings us to the next statement of the *S.G.* that expands it further: namely, what it really means to say that God is Lord, and – for Newton – an absolute Lord. For him, it means that there is no way of getting out of the Lord's power. Newton believed that he had come up with an unquestionable proof, that God is an absolute Lord of both the mechanical Cosmos and of us, the living creatures. God completely controls all things, at all times, in all places. That proof may have come from his concept of the world as a *phenomenon.*

Undoubtedly, Newton must have realized that we are in a strange bind: we have no way of finding out what this world *really* looks like and what the *real* essence of all things is. We are forever limited by our senses and our minds. We cannot transcend them. Here we run into a limitation that may be interpreted as somebody's supreme power. We may be getting ahead of ourselves, but let us quote from *S.G.*:

Videmus tantum corporum figuras & colores, audimus tantum sonos, tangimus tantum superficies externas, olfacimus odores solos, & gustamus sapores: intimas substantias nullo sensu, nulla actione reflexa cognoscimus. In bodies, we see only their figures and colours, we hear only the sounds, we touch only their outward surfaces, we smell only the smells, and taste the sa-

> vours; but their inward substances are not to be known, either by our senses, or by any reflex act of our minds.

That is the way things are. And, according to Newton, things are that way because Someone who rules with a supreme power deemed and executed them like this.

Let us add what will be expanded in point 22: whenever God is understood as a Lord and Ruler, it necessarily means a *personal* concept of God: He is a person, He has His own existence, which is independent of this world.

Let us say a few more words about the Greek word *pan.* In Greek, it means everything. And god Pan was also the object of Cudworth's interest in *T.I.S.*; among others, he explains here the meaning of the voice "The great god Pan is dead," heard – according to Plutarch – by the sailors. Cudworth also quotes Socrates' prayer to Pan, as well as several other authors, with a conclusion: Pan was a supreme Deity. Not a god of matter, but exactly the opposite: he was the intellectual principle, ruling everything, inspiring everything, penetrating everything. His *rule,* i.e. the idea that nature was divine, if completely absorbed in the divine mind, truly ended when Christianity took over the power over the world. Even Pan's name was taken over: the supreme God, and later also Christ, is called *Pantokrator.* The words "The great god Pan is dead" mean that the world is no longer divine, i.e., God stepped out of this world into transcendence.

The *Corpus Hermeticum* did not yet know of this change. It is not a Christian codex. Its god is still the ancient intellectual divine wisdom, all-penetrating, manifest in everything in this world. According to Cudworth's *T.I.S.*:

> We have now made it manifest, that according to the ancient Egyptian theology (from whence the Greekish and European was derived) there was one intellectual Deity, one mind or wisdom, which, as it did produce all things from itself, so doth contain and comprehend the whole, and is itself in a manner all things.[271]

C.H. Book X:

> Father of all Good, the Prince of all Order, and the Ruler of the seven Worlds.[272]

---

271 Ibid., p. 587.
272 Ibid., p. 589.

We conclude: obedience is a matter of relationship. Newton abandons the strictly rational and reflecting theology. As we mentioned in Chapter III, theology made a mistake in examining God as an object. This is untenable, because we can never step out of His dominion into independence. And especially for Newton, approaching God as an object was impossible.

7. God is a Living, Intelligent, and Powerful Being

> **Dominatio entis spiritualis deum constituit, vera verum, summa summum, ficta fictum. Et ex dominatione vera sequitur deum verum esse vivum, intelligentem & potentem; ex reliquis perfectionibus summum esse, vel summe perfectum.** It is the dominion of a spiritual being which constitutes a God; a true, supreme, or imaginary dominion makes a true, supreme, or imaginary God. And from his true dominion it follows that the true God is a Living, Intelligent, and Powerful Being; and, from his other perfections, that he is Supreme or most Perfect.

God as manifestation of all life is clearly expressed in the Hermetic Corpus. *T.I.S.*, Treatise XII (20):

> This whole world is intimately united to him, and observing the order and will of its father, hath the fullness of life in it; and there is nothing in it through eternity which does not live; for there neither is, nor hath been, nor shall be, any thing dead in the world.[273]

For the sake of completeness, we mention one of Newton's *Twelve Articles on God and Christ*:

> Article 1. There is one God the Father ever-living, omnipresent, omniscient, almighty, the maker of heaven and earth [...]

Those are standard attributes of God. They do not reveal anything special for our comparison of Newton and the Hermetic Corpus. Another point pertains to the divine power.

---

273 Ibid., p. 591.

8. Omnipotent

> **Aeternus est & infinitus, omnipotens & omnisciens, id est, durat ab aeterno in aeternum, & adest ab infinitio in infinitum: omnia regit [...]** He is Eternal and Infinite, Omnipotent and Omniscient; that is, his duration reaches from Eternity to Eternity; his presence from Infinity to Infinity; he governs all things [...]

T.I.S., Treatise VIII (10):

> Understand that the whole world is from God, and in God; for God is the beginning, comprehension and constitution of all things.[274]

Omnipotence is frequently mentioned by many authors. It is neither a specialty of Newton, nor of *C.H.* We will look at what it may mean for Newton in connection with the treatise *De possest*[275] by Nicolaus Cusanus.

Newton probably knew that treatise. It offers a new, so far unexplored comprehension of Newton's own work: Cusanus' *scientia* does not study the world, it studies concepts and relations between concepts. The objects of science are mutual interactions between concepts, and those interactions are twofold. Interactions between concepts are defined by physical laws and mathematical equations. But there is yet another kind of interaction: a particular concept may represent all its possible values within its class. Therefore, it also relates to infinity. And Cusanus claimed that a possibility cannot exist without supporting reality, and that reality is the spiritual, divine existence. Only God can warrant all the possibilities and all the values, because He is all the possibilities. Cusanus concludes: *scientia* studies the spiritual divine existence, therefore science is a part of theology. And God is omnipotent, because He rules over all of the possibilities, moreover, He *is*, at once, all possibilities and all realities. Only He is invariant. All else is just a phenomenon.

We feel that Cusanus' conclusions are valid for Newton and may be safely adopted for clarification of all of God's adjectives at this point.

---

274 Ibid., p. 590.

275 Cusanus: *De Possest*, Minneapolis: The Arthur J. Banning Press, 1986.

9. He knows things that can be done

> **Omnia cognoscit, qua fiunt aut fieri possunt.** [...] and [he] knows all things that are or can be done.

In Isaac Newton's *Twelve Articles on God and Christ* the same idea is expressed as follows:

> Article 4. The father is omniscient and has all knowledge originally in his own breast...

Here is another step toward phenomenalism: God knows all that is, therefore all that creates the phenomena of this world, but, being the Lord, He also knows and controls all that is not yet, and is no more, and what may be done, and – rationally – is not at all. Although Newton does not say it so literally, it follows from the logic of the argument that before anything happens, it already exists, in its substance, in the dark, unseen part of the world, therefore in God Himself.

That passage "all that is [...] and what may be done" is the central matter of the hermetic writings, namely, the topic of the *crossing of horizons*. One of the characteristics of the Greek god Hermes was to guide the souls in stepping across all sorts of horizons: in learning, at initiations, and even in death. The present horizon, between what is and what is not, is the domain of the god Hermes, the alleged author of the codex.

This inconspicuous sentence permits us to look for the most profound reason for the scientific method: working with models rather than with realities. Newton's natural science does not work with objects of this world, but with abstractions such as a *mass point*, a thing that does not exist. Yet working with those nonexistent models enables us to reach remarkably accurate predictions of future events.

We believe that here is the true reason: Newton's God knows and controls all that is, and also all that is not.

And from his statements, both from those mentioned above and those below, it follows that the word *know* has a powerful meaning of unifying knowledge and being. That is Plato's fifth dimension, but we also find it in the Bible: when Adam *knew* Eve, it means that he was united with her. Usages analogical to the Bible may be anticipated: Newton was a prominent Biblical scholar. His God contains in Himself the knowledge of all that is and all that is not.

But that exactly corresponds to the hermetic codex. *T.I.S.*, Asclepius (1):

> **Nonne hoc dixi, Omnia unum esse, et unum omnia, utque quia in creatore fuerint omnia, antiquam creasset omnia? Nec immerito unus est dictus omnia, cuius membra sunt omnia [...]** Have we not already declared that all things are one, and one all things? Forasmuch as all things existed in the Creator, before they were made; neither is he improperly said to be all things, whose members all things are.[276]

There is a passage in the text that pertains to the previously mentioned immanence: identifying God with the world. However, it does not mean that God is identical with the set of all things, nor that He would be created by putting all things together. *T.I.S.*, Asclepius (2):

> **Idcirco non erant, quando nata non erant, sed in eo iam tunc erant, unde nasci habuerunt [...]** And yet at that very time were they in him, from whom they were afterwards produced [...][277]

Cudworth offers several quotes from the hermetic codex No. V and translates its title as *Invisible God is most manifest*. *T.I.S.*, Treatise V (7):

> For there is nothing in the whole world which he is not, he is both the things that are, and the things that are not; for the things that are he hath manifested; but the things that are not, he contains within himself.
> (9) For what shall I praise thee? For those things which thou hast made, or for those things thou hast hidden and concealed within thyself?[278]

We offer additional quotations from *C.H.* Book V, in addition to those from *T.I.S.*:

> And if thou wilt force me to say anything more boldly, it is his Essence to be pregnant, or great with all things, and to make them.
> And as without a Maker, it is impossible that anything should be made, so it is that he should not always be, and always be making all things in Heaven, in the Air, in the Earth, in the Deep, in the whole World, and in every part of the whole that is, or that is not.

276 Ralph Cudworth: *The True Intellectual System of the Universe*, pp. 587–8.
277 Ibid., p. 588.
278 Ibid., p. 589.

For there is nothing in the whole World, that is not himself both the things that are and the things that are not.
For the things that are, he hath made manifest; and the things that are not, he hath hid in himself.
Thou Art All Things, and there is Nothing Else Thou art not.
Thou Art Thou, All that is Made, and all that is not Made. [279]

10. He constitutes Duration and Space

Non est aeternitas & infinitas, sed aternus & infinitus; non est duratio & spatium, sed durat & adest. Durat semper, & adest ubique, & existendo semper & ubique, durationem & spatium constituit. Cum unaquaque spatii particula sit semper, & unumquodque durationis indivisibile momentum ubique, certe rerum omnium fabricator ac dominus non erit numquam, nusquam. He is not Eternity and Infinity, but Eternal and Infinite; he is not Duration and Space, but he endures and is present. He endures forever, and is every where present; and, by existing always and every where, he constitutes Duration and Space. Since every particle of Space is *always*, and every indivisible moment of Duration is *every where*, certainly the Maker and Lord of all things cannot be *never* and *no where*.

Almost the entire remainder of *S.G.* will now follow the ideas on phenomenalistic approach to the world.

This point describes how Newton arrived at the idea of absolute space and time as concepts that cannot be comprehended by our senses. Absolute space and time are necessary and logical bases for his mathematical formulation of natural science, and even more. It ties on to the preceding point, i.e. absolute space and time phenomenalistically actually *do not exist,* but the existence of the *non-existent* is guaranteed by God Himself. It is also a condition of the origin of what does exist. In the following, we shall show how Newton understood this absolute space and time as a basis for this world.

Parallels for the idea that the entire world is contained in God may also be found in *C.H.* The hermetic documents are ancient and therefore use different terminology, however, the parallels for such a view of the world are noticeable, and they are equal in their concept. *T.I.S.*, Asclepius (5):

---

279 John Everard: *The Divine Pymander in XVII books*. London, 1650.
See http://www.levity.com/alchemy/corpherm.html.

> **Hic sensibilis mundus receptaculum est omnium sensibilium specierum, qualitatum, vel corporum.** The sensible world is the receptacle of all forms, qualities and bodies.[280]

Notice that McGuire's words on Newton's idea of space, mentioned above in the Chapter of *De Gravitatione,* are almost identical with this passage from Asclepius:

"Space is eternal and uncreated and is the *receptacle* within which all created things come into being,"

T.I.S., Treatise III (6):

> The divinity is the whole mundane compages, or constitution; for nature is also placed in the deity.[281]

A slightly different passage in *T.I.S.*, Asclepius (10), tells us that divinity encompasses everything within itself:

> **Si totum animadvertes, vera ratione perdisces, mundum ipsum sensibilem, et quae in eo sunt omnia, a superiore illo mundo, quasi vestimento, esse contecta [...]** And if you will consider things after a right manner, you shall learn that this sensible world, and all the things herein, are covered all over with that superior world (or deity) as it were with a garment.[282]

In this statement, divinity is not a space in which everything is located, but a divine garment that envelops everything. Newton understood that concept as it follows: a divine veil can only be an envelope of phenomena. It is a matter of formulation whether all phenomena of the world are placed in a divine receptacle, or covered by a divine veil.

We have to realize that what Newton called absolute space and time (and Kant later interpreted as ways of perception), leads to further approach toward a secret (arcane) theology of *C.H.*: we can comprehend the world only in a given, *a priori* established ways, independent of us. We only comprehend the phenomena of this world. Using the poetic words of *Corpus Hermeticum*, we only perceive that *divine veil* of this world.

Likewise, we may find the idea in *C.H.* that there is nothing in this world, no place, no time, except the divine *extent*. *T.I.S.*, Treatise XII (21):

---

280 Ralph Cudworth: *The True Intellectual System of the Universe*, p. 588.

281 Ibid., p. 589.

282 Ibid., p. 589.

And in this universum there is nothing which he is not: wherefore there is neither magnitude, nor place, nor quality, nor figure, nor time about God, for he is all or the whole (but those things belong to parts).[283]

T.I.S., Asclepius (5):

Quae omnia sine Deo vegetari non possunt: Omnia enim Deus, et a Deo omnia et sine hoc, nec fuit aliquid, nec est, nec erit; omnia enim ab eo, et in ipso, et per ipsum [...]
All which cannot be vegetated and quickened without God, for God is all things, and all things are from God, and all things the effect of his will; and without God there neither was any thing, nor is, nor shall be [...][284]

At this point, there is absolute agreement between Newton and *C.H.* Through His existence, God lets time and space come about. In this world, everything is a manifestation of God's will. It holds true for the phenomenalistic character of this world as well.

11. Thinking substance of God

Partes dantur successiva in duratione, coexistentes in spatio, neutra in persona hominis seu principio ejus cogitante; & multo minus in substantia cogitante dei. Omnis homo, quatenus res sentiens, est unus & idem homo durante vita sua in omnibus & singulis sensuum organis. There are given successive parts in duration, co-existent parts in space, but neither the one nor the other in the person of a man, or his thinking principle; and much less can they be found in the thinking substance of God. Every man, so far as he is a thing that has perception, is one and the same man during his whole life, in all and each of his organs of sense.

This is one of the most obvious points of similarity between the idea of God found in Newton and in *C.H.* One of the principal ideas in the *C.H.* is dealing with God as a cosmic mind, encompassing the world. That mind is uniform, never interrupted in time, nor space, like a human being: no matter what he perceives, as long as he is alive he remains the same human being. The Greek text of the *C.H.* uses the word NOUS, translated as mind, reason, or spirit. Asclepius, extant only in Latin, uses

283 Ibid., p. 591.
284 Ibid., p. 588.

the word SENSUS. In the first treatise, called "Poimandres," (this name was sometimes used for the entire collection of the hermetic codex), that divine mind introduces itself as *ho tes authentias nous*, thus the *consciousnes of itself, a spirit of authenticity, a mind that controls itself.* From that divine mind, somehow the *kosmos aisthetos* develops, the world perceived by our senses. *T.I.S.*, Treatise XI (17):

> You may consider God in the same manner, as containing the whole world within himself, as his own conceptions and cogitations [...][285]

C.H., Book I – Poimandres:

> The Father of all things, the Mind being Life and Light [...][286]

C.H., Book V:

> Of the Matter,
> the most subtle and slender part is Air,
> of the Air the Soul,
> of the Soul the Mind,
> of the Mind God.[287]

And also *C.H.*, Treatise XI:

> The Mind, O Tat, is of the very Essence of God, if yet there be any Essence of God.
> What kind of Essence that is, he alone knows himself exactly.
> The Mind therefore is not cut off, or divided from the essentiality of God, but united as the light of the sun.[288]

We may wonder what Cudworth meant by the title of his opus. He was a theist. Theists believe in a philosophical God, who created the world according to his plan and does not care about it any more. That *True Intellectual System of the World* is God as a cosmic Mind: a cosmic Intellect that designed all and set up natural laws so that the Universe may function. Newton agrees. This is his Thinking Substance of God.

---

285 Ralph Cudworth: *The True Intellectual System of the Universe*, p. 590.
286 John Everard: *The Divine Pymander*, *see* http://www.levity.com/corpherm.html.
287 Ibid.
288 Ibid.

In this connection, let us mention a part of Newton's Query No. 31 of his *Opticks*. Here we find his term *Sensorium,* i.e. a divine consciousness:

> Wisdom and Skill of a powerful ever-living Agent, who being in all Places, is more able by his Will to move the Bodies within his boundless uniform Sensorium, and thereby to form and reform Parts of the Universe, than we are by our will to move the Parts of our own bodies. And yet we are not to consider the World as the Body of God, or the several Parts thereof, as the Parts of God. He is an uniform Being, void of Organs, Members or Parts, and they are his Creatures subordinate to him, and subservient to his Will; and he is no more the Soul of them, than the Soul of a Man is the Soul of the Species of Things carried through the Organs of Sense into the place of its Sensation, where it perceives them by means of its immediate Presence, without the Intervention of any third thing. The Organs of Sense are not for enabling the Soul to perceive the Species in its Sensorium, but only for conveying them thither; and God has no need of such Organs, he being every where present to the Things themselves.[289]

Newton presents his idea of God who is omnipresent, therefore He can be aware of everything immediately, whereas living creatures perceive via their senses, therefore in two steps:

> 1. The sense organs transport the soul toward objects, so that the soul may perceive them. Our senses help us perceive partially and sporadically what God perceives directly through His mind.
> 2. In the second step, the true recognition of phenomena, *sensuum et phantasiae perceptiones,* takes place. According to Newton, objects have been endowed (by God) with the ability to appear to us. This is already mentioned in his *De Gravitatione*:
> **In mentibus creatis possint excitare varias sensuum et phantasiæ perceptionis [...]**

Although not identical with Treatise V, we feel that the similarity is important. *C.H.* speaks about "manifestation in phantasy." What God perceives directly in His *sensorium*, we receive in the form of phenomena. God creates the phenomena. We perceive those phenomena in our "imagination". We do not perceive the *world as it is.*

289 Isaac Newton: *Opticks or, a Treatise of the Reflections, refractions, Inflections and Colours*, London: W. Innys at the Prince's Arms in St. Paul's Church-Yard, 1717, p. 379.

C.H., Book V:

> For it needeth not to be manifested, for it is always.
> And he maketh all other things manifest, being unmanifest as being always, and making other things manifest, he is not made manifest.
> Himself is not made, yet in fantasy he fantasieth all things, or in appearance he maketh them appear, for appearance is only of those things that are generated or made, for appearance is nothing but generation. [290]

Zdeněk Kratochvíl, translator of the text into Czech, writes about how we try to translate Greek word *phantasia*. It does not mean just our fantasy, or in Latin *imaginatio,* but, literally, the ability to appear. Therefore the passage from Treatise V tells us that God has the ability to present (to us) everything in (our) comprehension. There is a slight difference: according to Newton, objects themselves, being God's creations, have that ability. However, both agree that the world is "under a veil" and must be "unveiled" for us.

12. God is the same God

> **Deus est unus & idem deus semper & ubique.** God is the same God, always and everywhere.

There may be just one God. It follows logically from the preceding: if He is omnipresent, omnipotent, etc., He cannot share the place with someone else, for instance, who is likewise omnipresent and likewise omnipotent. The quotation from Opticks in paragraph 11 explains why He is only one, and therefore always the same. God does not have any organs, or different parts. All that belongs to created bodies and beings. He is a homogeneous substance of a kind of mind. Therefore, it is impossible that, e.g., God would be absent somewhere, and that something mentioned above would be invalid, so that, e.g., (for us) the phenomenological nature of the world would fail.

For us, it is difficult to maintain this attitude for a long time. Therefore we repeat: we do not perceive the world as it is, but as it appears to us. We perceive it through our senses, those senses supply us with information, and our imagination catches the manifested activity (according to Newton), or God himself (according to *C.H.*). God's sensorium cannot fail;

290 John Everard: *The Divine Pymander*, see http://www.levity.com/corpherm.html.

the world cannot fail to manifest itself according to those laws. In *C.H.*, there are a number of passages stressing God's omnipresence, sameness and omni-penetration. *T.I.S.*, Asclepius (4):

> **Solus deus ipse in se, et a se, et circum se [...]** God alone in himself, and from himself, and about himself [...][291]

T.I.S., Book V (8):

> He is both incorporeal and omnicorporeal, for there is nothing of any body which he is not.[292]

Additional clarification of this subject will be given in paragraph 16.

13. He is omnipresent substantially

> **Omnipraesens est non per virtutem solam, sed etiam per substantiam: nam virtus sine substantia subsistere non potest.** He is omnipresent, not virtually only, but also substantially; for virtue cannot subsist without substance.

This is one of the cardinal points that all previous scholars overlooked, either intentionally, or by failing to notice Newton's evident heterodoxy: for he claims here that God is immanent, that He is omnipresent in substance. Such an opinion is neither compatible with Christianity, nor Judaism, and would have serious consequences. Nature, according to Jews and Christians (and also modern man who accepts their stand) is separate from God. Nature may be plundered, animals and enemies may be killed, flowers plucked, etc., without any remorse that, according to the older religion, the divine nature may feel pain. Therefore he is wounding the omnipresent body of God, for this is the consequence of a belief in an immanent God. Such an attitude was still prevalent in Egypt of the 13th century B.C.: rock temples were built and stone was quarried with great respect, because cutting rock meant cutting the body of the supreme god Amun.[293]

Newton is here returning to times prior to monotheism, whose God is transcendent toward the world. His God is a true Lord, an ancient God penetrating, connecting and enlivening the entire universe with all that is in it. He even makes this world comprehensible to us.

291 Ralph Cudworth: *The True Intellectual System of the Universe*, p. 588.
292 Ibid., p. 589.
293 Joseph Davidovits: *Nové dějiny pyramid*, Olomouc: Fontána, 2006, p.101.

T.I.S., Book V (9):

> And for what cause shall I praise thee? Because I am my own, as having something proper and distinct from you? Thou are whatsoever I am; thou art whatsoever I do, or say, for thou art all things, and there is nothing which thou art not; thou art that which is made, and thou art that which is unmade.[294]

Here we already find a clear, sincere image of God and His power in everything, even in the person of the writer: he confesses that there is nothing in him that would not be divine, neither what he says, nor what he does. That is a conspicuous hermetic text, confessional and also somewhat pantheistic. Again, we find an allusion to things that are not (yet revealed). *T.I.S.*, Book XVI (24):

> Who is the Lord and maker and father, and bound of all things; and who being all things, is one, and being one, is all things; for the fulness of all things is one and in one.[295]

C.H. does not use metaphysical terms and, of course, does not use *transcendent* and *immanent* concepts. It uses a different language: God is all things and in all things. It is not a language of precise conceptual thinking. However, it is an elegant language, of remarkable profundity, which is absent in rational thinking. *C.H.* does not talk about God as an object: God is only alluded to, very respectfully. He is at the foundation of everything. Hermetic philosophy is a humble, modest thinking, unlike its objective rational counterpart.

However, speculations about God's substantial presence go beyond this point: they go deeper because of their phenomenalistic viewpoint. God's substantial presence may just as well mean that everything (what is accessible at all), is accessible to our senses.

14. In him are all things contained

> **Deus nihil patitur ex corporum motibus: illa nullam sentiunt resistentiam ex omnipraesentia dei.** In him are all things contained and moved; yet neither affects the other: God suffers nothing from the motion of bodies; bodies find no resistance from the omnipresence of God.

294 Ralph Cudworth: *The True Intellectual System of the Universe*, p. 589.

295 Ibid., p. 592.

The question of God's omnipresence is further elaborated; the quotation is obviously derived from the Biblical Acts 17,28: "For in him we live, and move, and have our being; as certain also of your own poets have said..." But we know that it is Paul's paraphrase of Greek poet Aratus, who was not a Christian. For the sake of illustration, let us compare it with another of Newton's texts, approximately contemporaneous with *S.G.*, namely the manuscript *Isaac Newton's Twelve Articles on God and Christ*:

> Article 5. The father is immoveable, no place being capable of becoming emptier or fuller of him than it is by the eternal necessity of nature. All other beings are moveable from place to place.

The idea of an immoveable God appears several times in *C.H.* We quote *T.I.S.*, *Asclepius* (4):

> **Isque sua firma stabilitas est; nec alicuius impulsu, nec loco moveri potest, cum in eo sint omnia, et in omnibus ipse est solus.** [A]nd himself is his own stability. Neither can he be moved or changed by the impulse of any thing, since all things are in him, and he alone is in all things [...][296]

This is a very important theme. Such an idea is diametrically opposite to that of a stoic divine life-giving *pneuma*. The word *pneuma* means breath, wind, storm, therefore something characterized by movement. Here we have the opposite. We should keep it in mind in connection with the writings of all the scholars who speculate that Newton combined stoicism with Neo-Platonism, that he dematerialized the concept of pneuma. Newton's own words refute such speculations.[297]

T.I.S., Asclepius (5):

> **Omnia enim ab eo, et in ipso, et per ipsum [...]** [B]ut all things are from him, and in him, and by him [...][298]

For comparison, Newton's text of *Twelve Articles*, Paragraph 12:

> Article 12. To us there is but one God the father of whom are all things and we of him [...]

296 Ralph Cudworth: *The True Intellectual System of the Universe*, p. 588.
297 For instance B. J. T. Dobbs, D. Snobelen etc.
298 Ralph Cudworth: *The True Intellectual System of the Universe*, p. 589.

For comparison with *C.H.* Book V:

> O Son, what a happy sight it were, at one instant, to see all these, that which is unmovable moved, and that which is hidden appear and be manifest.[299]

We have to call attention to a footnote No. 3 in *S.G.*, where Newton quotes a number of ancient authors to corroborate his claim that all things are in God. He quotes Pythagoras according to Cicero, Thales, Anaxagoras, Virgil, Philo, Aratus according to St. Paul, in addition to several passages from the Bible. We repeat them here for the sake of completeness. It seems that even Newton was forced to use the method of "one sentence is enough," and from time to time protect himself with the words of the Holy Writ, even when those words, as in the present context, were weak, perhaps with the exception of the words of the prophet Jeremiah. Newton refers to the following Biblical passages:

> John 14,2 In my Father's house are many mansions: if it were not so, I would have told you. I go to prepare a place for you.
> Deuteronomy 4,39 Know therefore this day, and consider it in thine heart, that the LORD he is God in heaven above, and upon the earth beneath: there is none else.
> Deuteronomy 10,14 Behold, the heaven and the heaven of heavens is the LORD'S thy God, the earth also, with all that therein is.
> Psalms 89,7–9 God is greatly to be feared in the assembly of the saints, and to be had in reverence of all them that are about him. O LORD God of hosts, who is a strong LORD like unto thee? or to thy faithfulness round about thee? Thou rulest the raging of the sea: when the waves thereof arise, thou stillest them.
> 1. Kings 8,27 But will God indeed dwell on the earth? behold, the heaven and heaven of heavens cannot contain thee; how much less this house that I have builded?
> Jeremiah 23,23–24 Am I a God at hand, saith the LORD, and not a God afar off? Can any hide himself in secret places that I shall not see him? saith the LORD. Do not I fill heaven and earth? saith the LORD.[300]

However, Newton had not mentioned Cudworth at all. He often skipped the principal inspirator of his thoughts. We have to repeatedly

299 John Everard: *The Divine Pymander*, see http://www.levity.com/corpherm.html.
300 According to King James Bible.

stress the point that the previous text of S.C., the *Classical Scholia,* clearly point to Cudworth's inspiration in its very concept; and develops its argument almost parallelly: Cudworth claimed that ancient civilisation knew and practiced true monotheism, while Newton showed that ancient science knew the principles of his natural science, such as heliocentrism and gravitation.

The last sentence of note No. 3 to *S.G.*, which we quote in its entirety, is a quintessence of Cudworth's monumental work, *The True Intellectual System*, claiming that ancient nations, traditionally believed to be polytheistic, had the idea of a supreme God and should not be labelled polytheistic:

> The Idolaters supposed the Sun, Moon, and Stars, the Souls of Men, and other parts of the world, to be parts of the supreme God, and therefore to be worshiped; but erroneously.
> Fingebant autem idololatra solem, lunam, & astra, animas hominium & alias mundi partes esse partes dei summi & ideo colendas sed falso.

For those who know Newton, this sentence contains a hidden, strong and fundamental attack against the Trinity, where parts of God are worshipped as God Himself. Newton clearly claims that to be an error. He prepares here another point on the phenomenal nature of the world.

15. The supreme God exists necessarily

> **Deum summum necessario existere in consesso est: Et eadem necessitate semper est & ubique.** 'Tis allowed by all that the supreme God exists necessarily; and by the same necessity he exists always and every where.

This is one of the places where Newton's text is unduly terse. He claims that it is "allowed by all [...] by the same necessity," but we have to keep in mind how strong his phenomenalistic approach was. For him, God's existence is manifest at any instant and any point of this world that runs as it must, and we comprehend it with the complicated process of perception. Newton was fully aware that a portion of our comprehension is done by our "soul," (today we would talk about a psychic *projection*), that colours and sounds do not have any objective reality, that they are products of our psyche, derived from particular impulses of the outside world. Newton did not agree with Berkeley that all is nothing more than a play of our mind, without any "material" basis. Newton was more complex: from a material substratum comes a superstructure of our perceiving and it all takes place according to divine will, plan and constant labour.

For a more profound understanding of this phenomenalistic concept of the world, we have the full version of this true initiation text *C.H.*, Treatise V, chapter II, *The Unrevealed God is Most Revealed.* The Czech translators Zdeněk Kratochvíl and Radek Chlup thoroughly understood the idea of the phenomenalistic nature of the World, and in both words of the title of their translation into Czech appears the term *phenomenon.* Other translators used other, less accurate terms, such as invisible or unperceivable. *C.H.*, Treatise V:

> ... for whatsoever is apparent, is generated or made; for it was made manifest, but that which is not manifest is ever. For it needeth not to be manifested, for it is always.[301]

16. He is all similar, all eye, all ear, all brain

> **Unde etiam totus est sui similis, totus oculus, totus auris, totus cerebrum, totus brachium, totus vis sentiendi, intelligendi, & agendi...** Whence also he is all similar, all eye, all ear, all brain, all arm, all power to perceive, to understand, and to act...

Let us recollect the text from *Opticks,* which presents the idea of uniformity slightly differently:

> He is an uniform Being, void of Organs, Members or Parts, and they are his Creatures subordinate to him, and subservient to his Will.[302]

And we also have to keep in mind the divine "sensorium:"

> [He is] able by his Will to move the Bodies within his boundless uniform Sensorium [...][303]

It is an addition to point 11 about the thinking nature of God.

At this point, there is another argument against the Trinity: if God is uniform, homogenous, He could hardly be internally structured and

301 John Everard: *The Divine Pymander*, see http://www.levity.com/corpherm.html.

302 Isaac Newton: *Opticks*, p. 379.

303 Ibid., p. 379.

cannot contain three divine persons within Himself. In his opposition to the Trinity, Newton was adamant all his life; but as seen in this point, his arguments were often so delicate that they left an unprepared reader with the feeling that nothing had been proven.

McGuire erroneously claimed that Newton's "God is a structuring structure." Newton's God can never be a structure: He is patently uniform.

17. In a manner unknown to us

> **Sed more minime humano, more minime corporeo, more nobis prorsus incognito. Ut cacus non habet ideam colorum, sic nos ideam non habemus modorum, quibus deus sapientissimus sentit & intelligit omnia.** But in a manner not at all human, in a manner not at all corporeal, in a manner utterly unknown to us. As a blind man has no idea of colours, so have we no idea of the manner by which the all-wise God perceives and understands all things.

This is an interesting point: unlike in previous sentences, where Newton counts God's qualities and offers his own views of how God's world works, here he suddenly becomes negative, confesses his ignorance of what God looks like, what He is made of, how He perceives etc. and he will continue his negative list in the following paragraphs. We have not found any statements of this kind in *C.H.* Admitting one's own shortcomings is obviously beyond the possibilities of the sapiential text of the *C.H.*

18. He is void of body

> **Corpore omni & figura corporea prorsus destituitur, ideoque videri non potest, nec audiri, nec tangi, nec sub specie rei alicujus corporei coli debet. Ideas habemus attributorum ejus, sed quid sit rei alicujus substantia minime cognoscimus.** He is utterly void of all body and bodily figure, and can therefore neither be seen, nor heard, not touched; nor ought he to be worshipped under the representation of any corporeal thing.

For the sake of clarification, let us mention Article 2 from the *Twelve Articles:*

> Article 2. The father is the invisible God whom no eye has seen or can see. All other beings are sometimes visible.

As a matter of fact, starting with paragraph 9, all of Newton's text is interested in God as a planner, creator and "observer" of the phenomenalistic function of this world. Its content is a kind of analogy to paragraph 11: there, He had the nature of mind; here, He does not have a body and shape. *T.I.S.*, Treatise V (7):

> For there is nothing in the whole world which he is not, he is both the things that are, and the things that are not; for the things that are he hath manifested; but the things that are not, he contains within himself.[304]

T.I.S., Treatise XI (18):

> Is God invisible? Speak worthily of him, for who is more manifest than he? for this very reason did he make all things, that you mightest see him through all things: this is the virtue and goodness of the Deity, to be seen through all things. The mind is seen in thinking, but God in working or making.[305]

C.H., Book V:

> But do thou contemplate in thy Mind, how that which to many seems hidden and unmanifest, may be most manifest unto thee. For it were not all, if it were apparent, for whatsoever is apparent, is generated or made; for it was made manifest, but that which is not manifest is ever.
> For it needeth not to be manifested, for it is always. And he maketh all other things manifest, being unmanifest as being always, and making other things manifest, he is not made manifest.
> Himself is not made, yet in fantasy he fantasieth all things [...][306]

That is a complete manifest of phenomenalism. God presents (to us) everything in our imagination. He, invisible, creates – *C.H.* says *gives birth to* – our perceptions of the world, He makes the manifestation of this world in our imagination. *C.H.* Book V:

> Himself is not made, yet in fantasy he fantasieth all things, or in appearance he maketh them appear, for appearance is only of those things that are generated or made, for appearance is nothing but generation.[307]

304 Ralph Cudworth: *The True Intellectual System of the Universe*, p. 589.
305 Ibid., pp. 590–1.
306 John Everard: *The Divine Pymander*, see http://www.levity.com/corpherm.html.
307 Ibid.

Perceptions, "images in imaginations," are born in our sense organs in cooperation with our mind, which helps us to comprehend them. Those processes were later studied by Immanuel Kant, but we have to remember that it was Newton, who was the great inspiration for Kant; and we are gradually discovering that similar inspiration for Newton may have come from the Hermetic Corpus. *C.H.* Book V:

> But he is that One, that is not made nor generated, is also unapparent and unmanifest. But making all things appear, he appeareth in all and by all; but especially he is manifested to or in those things wherein himself listeth.[308]

That sentence may have flattered Newton. He was convinced that he was God's chosen one, a human to whom, in the words of the *Corpus Hermeticum*, God revealed Himself.

C.H. Book V:

> This is he that is most manifest; this is he that is to be seen by the Mind; this is he that is visible to the eye; this is he that hath no body; and this is he that hath many bodies, rather there is nothing of any body, which is not He.[309]

19. Much less have we any idea of the substance of God

> **Videmus tantum corporum figuras & colores, audimus tantum sonos, tangimus tantum superficies externas, olfacimus odores solos, & gustamus sapores: intimas substantias nullo sensu, nulla actione reflexa cognoscimus; & multo minus ideam habemus substantia dei.** We have ideas of his attributes, but what the real substance of anything is we know not. In bodies, we see only their figures and colours, we hear only the sounds, we touch only their outward surfaces, we smell only the smells, and taste the savours; but their inward substances are not to be known, either by our senses, or by any reflex act of our minds; much less then have we any idea of the substance of God.

At times, Newton permits himself some inconsistencies: e.g., he says that God is substantively everywhere (item 13), yet he admits that he does not know what that substance may be. Strictly speaking, that may be logically incorrect: if he does not know the substance, how can he be so sure that it is everywhere? Yet he is sure. It follows from the above

308 Ibid.

309 Ibid.

characteristics of God as a guarantor of the phenomenalistic nature of the world, a nature that cannot fail. It will also follow from point 20.

C.H., Book XI:

What kind of Essence that is, he alone knows himself exactly.[310]

20. We know him by his excellent contrivances of things

Hunc cognoscimus solummodo per proprietates ejus & attributa, & per sapientissimas & optimas rerum structuras & causas finales, & admiramur ob perfectiones; veneramur autem & colimus ob dominium. Colimus enim ut servi [...] We know him only by his most wise and excellent contrivances of things, and final causes; we admire him for his perfections; but we reverence and adore him on account of his dominion. For we adore him as his servants [...]

At the end of his treatise on God in *S.G.*, Newton returns to where the treatise started and offers another, even more condensed recapitulation of what he said above. He again uses superlatives: in point 1, he talked about the most elegant system; here he talks about the most wise and most perfect arrangement. He only points toward God. He does not speak of Him as an object.

He twice mentions final causes in that text. It is not clear, what was the final cause of creation for Newton. In that respect, *C.H.* is truly an important source of ideas.

C.H. Book XII, His Crater or Monas:

For Man became the spectator of the Works of God, and wondered, and acknowledged the Maker.[311]

This sentence is the quintessence of gnosis: the realization that brings man back to God. The purpose of creation is man's recognition, and the purpose of recognition is man's return to God. Kratochvíl writes: "Man 'was amazed and recognized': thus he opened his thinking. He realized his mind's faculty to partake in his Creator."[312]

This *opening of the mind*, this newly discovered recognition of the world, this initiation of the process of thinking is a most important

310 Ibid.

311 Ibid.

312 Zdeněk Kratochvíl: *Prolínání světů*, p. 74.

instant. To this day, it carries a great mystery. Not only as a memory of a young, fresh world, of a mankind not yet tired, but still hoping. It is a moment when mankind rose from the overwhelming myth, a moment when reason was budding and getting into equilibrium with emotion. That equilibrium is the balanced state of man. We admire it. To this day, it inspires us in the greatest prophets of the Old Testament, Isaiah and Jeremiah, as well as in the hermetic codex. An ideal human being may be understood as the final cause of God's creation:

C.H., Book XII, His Crater or Monas:

> For being Good, he hath dedicated that name unto himself alone. But he would also adorn the Earth, but with the Ornament of a Divine Body.
> And he sent Man an Immortal and a Mortal wight.[313]

Here we come across a great theme to which, however, *S.G.* does not return: i.e., the theme of man's salvation.

For better understanding, let us quote *C.H.*, Book V:

> I would it were possible for thee, O my Son, to have wings, and to fly into the Air, and being taken up in the midst, between Heaven and Earth, to see the stability of the Earth, the fluidness of the Sea, the courses of the Rivers, the largeness of the Air, the sharpness or swiftness of the Fire, the motion of the Stars; and the speediness of the Heaven, by which it goeth round about all these [...] O Son, what a happy sight it were, at one instant, to see all these, that which is unmovable moved, and that which is hidden appear and be manifest [...] O Son, how Man is made and framed in the Womb; and examine diligently the skill and cunning of the Workman, and learn who it was that wrought and fashioned the beautiful and Divine shape of Man [...] who set to outward view the more honourable parts and hid the filthy ones? [314]

There is absolute agreement between *S.G.* and *C.H.* on this point.

21. God without dominion is nothing else but Fate and Nature

Deus sine dominio, providentia, & causis finalibus nihil aliud est quam fatum & natura. A caca necessitate metaphysica, qua utique eadem est semper & ubique, nulla oritur rerum variatio. Tota rerum conditarum

313 John Everard: *The Divine Pymander*, see http://www.levity.com/corpherm.html.
314 Ibid.

pro locis ac temporibus diversitas, ab ideis & voluntate entis necessario existentis solummodo oriri potuit. God without dominion, providence, and final causes, is nothing else but Fate and Nature. Blind metaphysical necessity, which is certainly the same always and every where, could produce no variety of things. All that diversity of natural things which we find, suited to different times and places, could arise from nothing but the ideas and will of a Being necessarily existing.

In Francesco Patrizi's book of Latin translations, there is a treatise called *De providentia & Fato. Ex libris ad Ammonem.* It deals with the problem of blind fate. It does not have a number, because, as was discovered by later scholars, it was taken from quotations of hermetic sources by Stobaeus. That treatise shows a conspicuous agreement with Newton's ideas in *S.G.*

Omnia vero fiunt natura & fato. Et non est locus desertus a providentia. Providentia vero est per se perfecta ratio, supercoelestis Dei. Duae autem sunt ab ea notae potentiae, necessitas, & fatum. Fatum autem ministrum est providentiae & necessitatis.[315]

Indeed, everything happens due to its nature and fate. And there is no place where providence would be absent. Providence itself is truly a perfect reason of God above heavens. We recognize its two powers: necessity and fate. But fate is a servant to providence and necessity.

22. He is a person

Dicitur autem deus per allegoriam videre, audire, loqui, ridere, amare, odio habere, cupere, dare, accipere, gaudere, irasci, pugnare, fabricare, condere, construere. Nam sermo omnis de deo a rebus humanis per similitudinem aliquam desumitur, non perfectam quidem, sed aliqualem tamen. But, by way of allegory, God is said to see, to speak, to laugh, to love, to hate, to desire, to give, to receive, to rejoice, to be angry, to fight, to frame, to work, to build. For all our notions of God are taken from the ways of mankind, by a certain similitude which, though not perfect, has some likeness, however.

It follows from the above list that Newton comprehends God as a person. That means that God is not just immanent in the world, but he

315 Francisco Patrizi: *Magia Philosophica*, Hamburgi, 1593, p. 220.

transcends the world by His own existence, too. That puts to rest any idea about Newton being a pantheist like Bruno, whose God is completely exhausted by the existence of the world and does not transcend the world. At this important point, Newton again agrees with many other references to *C.H.* and their confession of a personal God as father or Lord. See, e.g., quotations no. 3, 8, 9, 20, 22, and 24.

23. To discourse of whom from the appearances of things

> **Et hac de deo, de quo utique ex phaenomenis disserere, ad philosophiam naturalem pertinet.** And thus much concerning God; to discourse of whom from the appearances of things, does certainly belong to Natural Philosophy.

This sentence is the apex of the entire treatise about God in *S.G.* All the arguments concerning Newton's phenomenalistic approach toward the world come to a head. We take that sentence as a confirmation of legitimacy of our parallel study of *S.G.* and *C.H.* A rough translation might sound: To deal about God because things appear to be. According to Newton, God may be proven and demonstrated by His activity. The preposition *ex* means here not only a location, but also a cause, purpose or condition.

Summary of Chapter VI.

Our goal was to discover, through a detailed analysis, to what extent Newton's concept of God agrees with the supreme divine being of the hermetic codex.

We conclude: Newton's God is incongruous. Simultaneously, He is *transcendent* – i.e. personal, having His own existence independent of the world, for otherwise, He could not rule the world. And, also, in His own way, He is *immanent*. He is everywhere, *substantially* present in everything. He created an imperceptible *absolute* space, containing – as in a receptacle – the entire cosmos. This space is God's "sensorium", His indivisible complex of mind and sensory organs. To us, the world is accessible only through its appearances, which, in turn, is God's activity which at all times and places makes our life possible in all its ramifications.

The God of the *C.H.* is likewise characterized by a paradox and, in His fundamental characteristics, agrees with Newton's ideas. He is simultaneously *transcendent* and *immanent*. The way of thinking of the hermetic writings does not have Newton's precision of thought and expression. If it did, perhaps the agreement would be perfect. In any case, we believe that the similarity between the unrevealed, yet all revealing God of Treatise V, and Newton's phenomenalistic concept of natural philosophy has been documented.

In order to identify Newton's heterodoxy, we clarify here some basic theological concepts:

Theism is a common modern monotheism where God is *transcendent* in relation to the world. That means that He is beyond this world, beyond our experience. He is separated from His creation. As mentioned above, theism originated as a consequence of the progressing rationalization of thinking and de-deification of the world.

Pantheism means that God is identical with the world. He is inseparable and indistinguishable from the world and does not exceed the world. A change of the world means a change of God. Therefore, God is a kind of driving mechanism, or spirit of the world (*anima mundi*). He cannot be beseeched. Among the best-known pantheists were, e.g., Giordano Bruno and Baruch Spinoza. God is entirely *immanent* (as opposed to transcendent).

Panentheism claims that the world is in God, God penetrates the world, the world with its multiplicity of individuals is contained in God as in a supreme unity, and God acts through them. Therefore, God is not separated from the world, as He is in theism. Among the proponents of panentheism, we find Malebranche.

Deism, more precisely panentheistic deism, originated toward the end of the 17th century, precisely during Newton's lifetime. The above-mentioned Spencer and Cudworth are sometimes labelled as Neo-Platonists, other times as deists. The God of the deists is a philosophical God. He once created the world, but does not interfere with the world any more, although He still somehow makes it run. Deism rejected the category of Revelation and of a personal relationship with God. It sought a natural theology that would rest only on an abstract idea of God and required only moral commitments from man. Deists are often characterized as rationalists with a heart unconsciously thirsting for a religion of empathy. Besides those Cambridge deists mentioned above, let us mention Voltaire in France and Paine in America.

Paganism is usually viewed as an antithesis to monotheism. However, as shown by Cudworth's penetrating analysis, polytheism is not its main sign, because even here is one supreme deity (power) who may also be worshipped as a divine person. Polytheism exists only in an exoteric paganism, it's actually kind of concession to the plebeian imagination.

It may seem that Newton stood close to deism; and, sometimes, he is believed to have been a deist. However, there is an important difference: Newton's God is a person with whom one may, should, and must have a personal relationship.

Corpus hermeticum II.

Jan Assmann, who analyzed the main work of Ralph Cudworth, is also the author of several books about Old Egypt and its ancient religion. For many years, he was a professor of Egyptology at the University in Heidelberg and since 1978 he had been conducting excavations at Luxor in Upper Egypt. At present, he is probably the best authority on ancient Egyptian religion.

Assmann describes in detail the evolution of Egyptian religion from the earliest times. He uses the term *the dimension of God's presence,* and studies the difference between *explicit* and *implicit* theology. Those terms correspond to Cudworth's *common* and *arcane theology*. Although they seem to be distant from our topic, those Egyptological details deserve further mention.

At first, there were three dimensions of Old Egyptian implicit theology. They derived from the cultic life. Assmann defines them as the language dimension (gods' names, mythus), the spatial dimension (gods of cities represented in their statues and temples), and the cosmic dimension (cosmology as a *constellation* of the gods).

Those three original canonical dimensions were gradually, along with increasing human rationality, getting into an ever-deeper conflict with the fourth dimension, a dimension of *God's closeness* and direct experience. The subconscious feeling of a supreme God, yet without any cultic manifestation, grew into accepting a hidden divine being toward whom one can feel a deep personal piety instead of the old, mechanical rituals. This new kind of piety dates from the queen Hatshepsut (1473–1458 BC) of the XVIII dynasty.

Eventually, that process brought about violent and generally unfortunate reform of Akhenaten (1352–1336 BC), who tried to change the entire framework of Egyptian religion. Not only did he try to remove its *constellative* aspect, (the cooperation of several gods in running the world), but, unfortunately, also to stop everybody's contact with the supreme God. Undoubtedly, here we find one of the reasons for his failure.

Assmann thoroughly compares the traditional supreme god Amun with Akhenaten's Aton. Our romantic ideas about an enlightened pharaoh, whose monotheism was ahead of his time, start to break down. Only the emperor and his wife, Nefertiti, had access to the new God Aton. The rest did not worship Aton, but they worshipped the pharaoh and his wife. Aton was a silent cosmic God from whom the emperor did not expect any response to his invocations. He was not a personal god; therefore, from the vantage point of the history of religion it was a regressive step.

After the death of this heretical pharaoh, religion returned to the traditional gods with Amun at their head. However, some changes introduced by that rigorous revolution could no longer be erased. In particular, it became clear that the world would run without the constellative function of many gods. Therefore, Akhenaten, and not the Bible, was the one who de-deified nature and opened an entirely new way of studying the World. Assmann sees Akhenaten's ideas as a kind of *Natural Philosophy* that interprets all phenomena of the world and life as the action of some elementary divine power. "It replaces the radically rejected constellative mythology [...] with a religious phenomenology based on visible divine works [...] The horizon frees itself from the captivity of mythical images and reaches an almost natural-philosophical, while still theological, comprehension of the cosmos. A cognitive revolution took place [...]"[316]

In Akhenaten's texts, we find the first clearly phenomenalistic interpretation of the world. It is sunlight that makes the world visible, it is Sun (light) which creates the visible world.

With the return of the traditional pantheon, a thorough change of the entire nature of religion took place. Originating with Queen Hatshepsut, religion became confessional. That change culminated during the reign of Ramesses II (1279–1213 BC). The most diversified and most profound texts illustrating this change originated in that period. The theological arguments brought about an unusual concept of God; and this concept may have become the model for the supreme God of the hermetic codex. And, as such, it may have also inspired modern hermetics.

The basic pattern and trademark of this new confessional theology is a paradox. Various aspects of God are expressed by sharp contradictions whose "not only – but also" points toward a *transcendental* God, i.e. a rational comprehension of an incomprehensible substance, and at the same time an *immanent* God.

---

316 Jan Assmann: *Egypt*, Praha: Oikoymenh, 2002, p. 246.

The *personal closeness*, the fourth dimension of divine interaction with man, is described in the Ramesses texts as an action of the subject's free will, therefore as a helper in need, a savior, but also as a judge and avenger. Inscriptions on the stelae list individuals who have "found" God. Here we encounter a personal experience of God's close presence. Of course, that divine will must meet with the "God of life" In history, divine power manifests itself as blessing and wrath, and at the same time in the life-giving elements of air, water, and light.

Such a cosmic theology (cosmotheism) culminates in the magnificent vision of World-God, where the entire cosmos is a visible body of God. However, such an almost pantheistic concept of God is aware of the conflict between the cosmos and history, i.e. simultaneous incarnation and regulation. God is life not only in the sense of the elemental cosmic powers (sun, moon, water, wind), which are His body, but also in the sense of cause and fulfilment, success and prosperity, which He grants according to His will, but keeps in mind those who sincerely worship Him and those who do not.

Lord over life,
He who gives to whom He wants to give. The world is under His supervision.[317]

The God of Ramesses combines two contradictory aspects: the all-enlivening cosmic principle and free will. This Lord of the world is mysteriously dual. He is a personal God:

As a spectator He is far away,
As a listener He is near at hand.[318]
He who foretells the future for millions of years,
Eternity before His face
Is like yesterday. [319]

Here is an idea of a God whose omniscience encompasses both the temporal and spatial dimension of this world. He is a cosmic God, too:

Thy eyes are the sun and the moon,
Thy head is the earth,
Thy feet are the underworld.[320]

317 Neschons 23–24 = J. Assmann, ÄHG, No. 131, l. 58 f.
318 Berlin 3049, VIII, 4 = J. Assmann, ÄHG, No. 127 B, l. 34 f.
319 J. Assmann, ÄHG, č. 127 B, l. 81 f.
320 J. Assmann, STG, No. 88, l. 12–15.

In those texts, Assmann recognizes a remarkable unity of theology, mentality, and a way of life. *The pantheistic theology of transcendence* developed a new nomenclature that, among others, offered to accept the multiplicity of gods as mere reflections of a hidden divine unity, ("Amun" means hidden), as splinters of a supreme God. This hidden God gradually lost His name, had no form. No picture could describe Him, man was not able to name Him.

This was a necessary introduction to Assmann's conclusions pertaining to the Hermetic Corpus and Ralph Cudworth's *The True Intellectual System of the Universe*. Assmann shows that Cudworth – unlike Casaubon – believed in the authenticity of the Hermetic Corpus. More than a hundred years after Cudworth, the hieroglyphs were deciphered, and another two centuries passed before an analysis of Egyptian literary documents was completed. At long last, we may solve the old argument between Casaubon and Cudworth. Assmann concludes his chapter on Cudworth as follows: "Erst jetzt sind wir in der Lage, jene Monumente zu studieren und jene Inschriften zu lesen, die Cudworth vergeblich suchte. Die hieroglyphischen texte bestätigen Cudworths Intuition in einer Weise, wie er es nicht besser hätte wünschen können."[321]

That means that modern Egyptology and modern studies of old Egyptian religion admit that the paradoxical concept of the immanent and transcendent God of the Hermetic Corpus probably agrees with the Ramessesian idea. And that idea in turn developed due to the failure of Akhenaten's religious reform. Assmann even found a hieroglyphic analogue to the Greek concept of *to hen kai pan,* "one and many." Assmann reads it as "the one who makes himself into millions," or "one whose body is millions." In old Egyptian sources, the concept of "everything" and "completeness" is expressed as "millions." Of course, it does not mean literally such numbers, but a number beyond counting. Moreover, etymologically, that word is related to the Egyptian term for "eternity." According to Egyptian ideas – unlike the Biblical one where everything came from a pre-existing, transcendental One – everything emanates from the One all the time, everywhere, and beyond that One, nothing else exists.

Those discoveries open further opportunities for research, beyond the range of the present study. Until now, there has been an open question of what is original in the works of Greek philosophers, and what has

321 Finally, we are in the position to study those documents and read those writings Cudworth hoped to find in vain. Those documents support his intuition beyond his highest hopes.

been adopted from older sources. Those older sources are Egypt and its mysteria, i.e. Cudworth's "arcane theology." It was the first theology that recognized the phenomenalistic existence of this world. It may also be suggested by the hieroglyph 𓁹 which has a double meaning: to see and to create.[322] Perhaps it means that by seeing we create the world.

We close this chapter with Assmann's assertion that, with respect to the concept of God, the Hermetic Corpus is genuine. The supreme God of the Hermetic Corpus corresponds to the concept of God of the Ramesses period, about the 13th century BC.

322 Raymond O. Faulkner: *A Concise Dictionary of Middle Egyptian*, Oxford: Griffith Institute, 1988, pp. 25–6.

VII. Conclusions

Slowly with the rolling years the obvious (so often the last revelation of analytic study) has been discovered...
J. R. R. Tolkien[323]

Let us now summarize the results of our analysis.

As mentioned previously, Newton's image is currently in a stage of change. Our research therefore focused on the problem of Newton's concept of God. Of course, we studied all aspects of current research on Newton and used the results of prominent contemporary authors, particularly B. J. T. Dobbs and K. Figala in the field of alchemy; F. Manuel, R. Iliffe, S. Snobelen and F. Wagner in Newton's theology; and J. McGuire, P. Rattansi, A. Hall, and many others in natural sciences. J. Harrison's catalogue of Newton's Library was of great help. However, our own research owes its principal debt to the German specialist on Egyptian religion and culture, Jan Assmann. He introduced us to the problems of *prisca sapientia* and its influence upon Cambridge deists of the 17th century. We also took advantage of his old Egyptian studies, such as the natural philosophy (Naturphilosophie) of Akhenaten, the paradoxical comprehension of the divine being at the times of Ramesses II and the revolutionary ideas about the authenticity of the Corpus Hermeticum. Last but not least, Jan Assmann gave us materials on Newton the historian. Among them were some facts on the mythical Jewish exodus and the mythos of the historical Moses. Of Czech authors, let us mention Z. Kratochvíl and his Czech translation of the Greek hermetic literature, as well as Zdeněk Neubauer, who inspired us with his profound ideas about alchemy and hermetism.

323 J. R. R. Tolkien: *The monsters and the critics, and other essays*, Michigan university: Allen & Unwin, 1983, p. 9.

Chapter II about Newton's place in the spiritual and philosophical background of his times is intentionally and selectively focusing on points of our interest. Among Newton's sources of inspiration, we have concentrated on Francisco Patrizi and Ralph Cudworth. Their work was of enormous value for our own analysis of the *Scholium Generale* and the Hermetic Corpus.

Our analysis required a thorough understanding of all of Newton's activities and his complex intellectual life. Those are discussed in Chapters III, IV and V and will not be repeated. Instead, we shall summarize our findings as follows:

Although Newton's Christian faith was profound and he was a thorough Biblical scholar, many of his activities were at variance with Christianity. Those activities got the upper hand, while Newton himself was not aware of it.

It was his dependence on the *prisca sapientia* and his lifelong practice of alchemy that enabled him to come up with an entirely new concept of nature, based on ancient ideas.

While our own research was focused on other questions, we feel that some of our results contributed to the general comprehension of various hidden origins of modern science. We started with the quotation from John M. Keynes who calls Newton the last magus of the ancient times. We have shown that Newton's connection with his predecessors, even with some of the very distant ones, should not be underestimated. Previous authors have already noticed that Newton believed in ancient wisdom, *prisca sapientia,* which was subsequently lost. Newton felt that he was the one who had been called to rediscover the ancient knowledge. In a way, he was an esoteric. Not only because of his interest in alchemy and his knowledge of hermetic writings, but also because of his intimate inclination toward searching the hidden, secret, mysterious, complex riddles. Even in his Biblical studies he was of the opinion that knowledge of the simple principles of faith, the *milk for the infants,* is enough for the salvation of the soul, but a different spiritual food is needed by the grown-up, strong men. For this reason, his writings are addressed exclusively to the class of readers in whom he expected a ripe spirit, strength, and an invincible quest for understanding the most complicated problems. Even nowadays, more than three centuries after their publication, his *Principia* is difficult reading for a non-specialist; and his *Scholium Generale* reads like an initiation text which assumes that the reader holds a key (or, perhaps, that he will enjoy the favours of the divine Hermes, Lord of All Knowledge.)

Newton followed a trend which started in the Renaissance: i.e., the effort to purify Christianity, free it of confessional schism and enrich it with *prisca sapientia,* the heritage of the primeval revelation of Divine Wisdom to Noah. It seems to us that Newton tried to encompass the whole world, and (up to a point) all the accessible knowledge of his time. He solved that problem by dividing his work into three spheres:

1. The sphere of the non-living matter, obedient to strict mathematical language, i.e. of natural science.
2. The sphere of the living matter, i.e. of alchemy.
3. The sphere of the spirit, i.e. of theology and history.

Those spheres, dealt and summarized in their particular chapters, were united in the person of Isaac Newton. Always trying to conquer chaos, satisfy (often unsuccessfully) the human quest for understanding of this world, and expressing its *sense* in a *rational* way. Perhaps we may claim that Newton honestly tried to put rational thinking in the service of value judgement, because all his activity was directed toward his duties toward the Absolute Value, toward God.

We attempted to confront Newton's ideas in the *Scholium Generale* with the hermetic quotations from the books of Ralph Cudworth and Francisco Patrizi. We found similarities between hermetic writings and Newton in the concept of God, which, for Newton, was of fundamental importance. Both Newton's God and the hermetic God are paradoxical, because they are simultaneously *transcendent* and *immanent.* At the time of its origin, i.e. in the second century AD, the Hermetic Corpus incorporated the concept of God developed and formulated at the time of pharaoh Ramesses II, about 1500 years earlier. It then took another 1500 years before that concept reached Newton. Therefore, Newton developed his rational, mathematical natural philosophy on this 3000 years old concept.

Our conclusions are supported by the quotation from the well-known scholar James E. McGuire:

> It seems that Newton never fully resolved two distinct pictures of how God relates to the world: the Arian, or something similar to it, in which the high God is transcendent and works in nature through an intermediary, the cosmic Christ; or the God of dominion of the *Classical scholia* who is directly present and active in creation [...][324]

324 McGuire, James E.: "The Fate of the Date: The Theology of Newton's *Principia* Revisited," in: *Rethinking the Scientific Revolution*, Margaret J. Osler (ed.), Cambridge: Cambridge University Press, 2000, p. 294.

McGuire thus sees the paradoxical God as Newton's error, or indecision. We, however, believe that McGuire overlooks the theological subtlety of the problem. This paradoxical God, simultaneously *transcendence* and *immanence* is not Newton's oversight. It is the very nature of Newton's God.

Newton built his natural philosophy on a very old image of God. While using contemporary scientific discoveries, experimental observations and modern mathematical methods, he left out the entire cultural development of the Western World and built his system on middle-Platonic *Corpus Hermeticum*, which some of his contemporaries deemed decadent, if not a complete falsification. Newton skipped the cream of Greek philosophy, medieval learning and scholasticism. He departed from Descartes. And, believing he was returning to an earlier, more authentic record, he seriously modified the message of both the Old and the New Testament. If Assmann's opinion pertaining to the age and authenticity of the concept of God of the hermetic Corpus are correct, then Newton's approach was well-founded.

Let us return to the main point of Chapter VI. Contemporary theological definition teaches that the substance of theology is God in His revelation. In the traditional Christian theology, that revelation is Jesus Christ and his mission. However, for Newton, Jesus was just a human being. His statement, taken from the Scholium Generale, "Et hac de deo, de quo utique **ex phaenomenis** disserere, ad philosophiam naturalem pertinet" leaves no doubt. All of Newton's work is theology, as long as we agree that Newton brought forth a non-traditional, cosmologic concept of divine revelation, independent of the historical Jesus, but manifested in the run of this world. This then agrees with the definition of theology as the science of divine revelation.

Newton was interested only in that kind of constant, eternal divine revelation manifested by the run of this world. Such a concept is alien to Christianity, but it is not new. We find it, e.g., in the Fifth Hermetic Tractate: *Quod immanifestus Deus manifestissimus est (Invisible God is most manifest).*

Newton thus tried to create an entirely new kind of theology, compatible with Hexameral literature, free of conflict between the Word of God and scientific findings. Such a *physicotheology,* according to McGuire, or *theophysics*, was a powerful apologetic tool, equipped with an unconquerable weapon: mathematics. In a sense, it was a theology of *liberation.* It freed man of old dogma by means of reason. Newton was victorious. Old dogmas fell. Nevertheless, Newton himself would have probably called it a Pyrrhic victory.

Newton himself would have certainly been critical of our conclusion, but we have to pronounce it anyway, because it logically follows from our study:

In good faith that he was returning to the oldest, most authentic source, Newton in fact created (or recreated) a kind of pagan theology.

Unlike their monotheistic counterparts, pagan religions, both ancient and modern, recognize a single, divine, life-giving power; they are an older, more primitive, less rational kind of religion. With increasing rationality there was an intention to remove the God from the material world: we may see, for instance, that both the Old Testament and Plato required a complete separation of God from the world. Assmann speculates that while paganism seeks to find a home in this world (Weltbeheimatung), the most rational monotheism seeks to conquer the world (Weltüberwindung). The true Christian finds in his faith no motivation to study how this world works. Newton, however, missed one of the principal tenets of rational thinking: the separation of God and world. Not only was Newton's world paganly divine: Newton studied the world because it was paganly divine. An orthodox monotheist could not be so interested in this world, because his interest is focused on the *other world*. That, of course, means that Newton's rational physics, and modern science derived from that physics, is taking an intellectual step backwards. Further, a sort of ignorance is hiding in the very foundations of modern science: Newton had not known that he was a pagan. Neither does modern science know that it studies divine *manifestation*. And even further. Newton introduced pagan cultus of the *number*, his *theophysics* is real *theocracy* through numbers and calculating. And this all-controlling god is, according to our results, the well-known god Thoth, Greek Hermes, Roman Mercurius.

Yet we should not be disappointed by recognizing that Newton was not who he thought he was. His *theophysics*, retained in his *Principia*, is still valid and useful, despite the fact that its author's original intention vanished. It had to: to be beyond scrutiny, it had to be methodically exact. It had to be divided into a realm of exact mathematical formulations of some phenomena, and a realm of speculative boundary conditions, values, that do not obey rules of reason. Those boundary speculations concerning God as an absolute value, correctly and methodically removed into the *Scholium Generale,* were easily removed onto a shelf of curiosities. What was left was a fruitful opus branching into new discoveries, but not what its author had in mind. The spiritual prerequisites of physics, found in *Scholium Generale* for the *Principia* and in the *Queries* to *Optics*, fell into oblivion. And the fate of the *Scholium* is reminiscent of the fate of its God, falling slowly into oblivion...

Modern natural science, according to our study, is the result of a pagan concept of God. That concept is its absolute necessity. The purpose of science is – or was, according to Newton – to study and describe God in all accessible realms of His doings. For He is the one who reveals Himself in all events, always in a particular way out of all possible ways hidden in His spiritual dimensions. Today, we know that this concept of science as description of the doings of a supreme God comes from the pharaoh Akhenaten and that a second coming of such a concept of science had to wait for Newton. We then conclude that European science does not rest on Christianity, but that it developed in spite of Christianity. And therefore the key to modern civilization does not build upon Christianity, neither on Judaism as Christianity's predecessor, but rather on an old Egyptian *cosmotheism,* communicated by the Hermetic Corpus.

These are big words. However, we feel that we have accumulated sufficient corroborative evidence for them. That, of course, opens many new questions: we shall let others answer those questions.

Newton himself thought that he was building on the ideas of Hermes Trismegistos and on Moses, whom he accepted as historical personages. We believe that the true historical personages on whom Newton built his science were pharaoh Akhenaten and pharaoh Ramesses II. Although Newton did not, and could not, know those two persons, perhaps – with some poetical license – we may insert them into his metaphor:

"I was standing on the shoulders of giants."

Zusammenfassung

Mehrere Jahrhunderte lang war Isaac Newton ein Symbol der Vernünftigkeit gewe-sen. Wenn aber in den dreißiger Jahren des 20. Jahrhunderts alle vorhandenen Manuskripte Newton's für Gelehrte zugänglich wurden, fing das Bild von Newton an sich zu ändern. Für einige weitere Jahrzehnte wurde Newtons umfangreiches Opus im Feld von Alchemie, Theologie und der Geschichte als eine Laune eines senilen, merkwürdigen Genies interpretiert. Nur vor kurzem ist die Idee ernsthaft betrachtet worden, daß Newtons wesentliche Arbeit in Naturwissenschaften inspiriert wurde, eben konditionierte, durch seine lebenslange systematische und zähe Arbeit in nicht-wissenschaftlichen Disziplinen. Zeitgenössische Newtonian Studien, den Vorteil einiger Jahrzehnte vorheriger Forschung habend, sind im Stande, der sich ändernden Perspektive. Eine Periode von Datenansammlung folgend, sehen wir ihre Auslegung von Experten in mehreren Disziplinen.

Diese These weist auf die gegenseitige Verbindung zwischen seiner wissenschaft-lichen und nicht-wissenschaftlichen Arbeit, auf den kulturellen und geistigen Trends seiner Zeiten und Quellen seiner Inspirationen, nachdem sie ein gesamtes Bild von Isaac Newton und seinem Leben präsentiert hat. Danach umfassen zwei Kapitel die erst nicht-wissenschaftlichen Interessen von Newton d.h. Theologie und Alchemie. Jene Kapitel umreißen die Trends von Gedanken, der den Pfad zu seiner neuen und originalen Vorstellung wissenschaftlicher Forschung öffnete.

Das nächste Kapitel behandelt die natürliche Philosophie von Newton. Es gibt eine ausführliche Analyse seines Manuskripts *De gravitatione*, ein unmittelbarer Vorläufer von seiner wesentlichen Arbeit, dem *Principia*. Es ist hier, wo wir vielleicht Newtons Trend der Idee folgen: seine Physik ist eigentlich eine *physico-Theologie*, weil es erforscht und die Welt als Manifestation ununterbrochener höherer Gewalt beschreibt.

Der Hauptteil dieser Abhandlung präsentiert dem *Principia* einen ausführlichen Vergleich vom Epilog, rief *Scholium Generale*, mit ausgewählten Zitaten von hermetischen Schriften, die von Raphael Cudworth in seinem *The True Intellectual System of the Universe* zitiert werden. Das wichtige Einverständnis zwischen jenen zwei Texten d.h. die gleiche Idee des Paradoxes von Gott sowohl transzendent, als auch immanent, verlangte einen Vergleich mit letzten Ergebnissen der Forschung, die das ägyptische hermetische Korpus betreffen. Da Newton seine Physik an Gott baute, der den hermetischen Schriften wiederum entspricht und jene Schriften dann die alte ägyptische Vorstellung von Gott weiter aufzeichnen, dann folgt es von Notwendigkeit, daß die moderne europäische Naturwissenschaft auf anderen Grundlagen steht, als wir dachten.

Literature

Alleau, René: *Aspekty tradiční alchymie*, Praha: Merkuryáš, 1993.

Alleau, René: *Hermes a dějiny věd*, Praha: Trigon, 1995.

Assmann, Jan: *Moses der Ägypter*, Frankfurt am Main: Fischer Taschenbuch Verlag, 2004.

Assmann, Jan: *Egypt*, Praha: Oikoymenh, 2002.

Assmann, Jan: *Egypt ve světle teorie kultury*, Praha: Oikoymenh, 1998.

Assmann, Jan: *Monotheismus und Kosmotheismus*, Heidelberg: Universitätsverlag C. Winter, 1993.

Assmann, Jan: *Die Mosaische Unterscheidung*, München und Wien: Carl Hanser Verlag, 2003.

Assmann, Jan: *Kultura a paměť*, Praha: Prostor, 2001.

Balabán, Milan: *Hebrejské člověkosloví*, Praha: Herrmann & synové, 1996.

Berman, Morris: *The Reenchantment of the World*, Ithaca: Cornell University Press, 1981.

Bible, version of King James, http://ebible.org/bible/kjv./.

Bor, D. Ž.: *Alchymická tvrz*, Praha: Trigon, 1992.

Bor, D. Ž.: *Napříč říší královského umění*, Praha: Trigon, 1995.

Bor, D. Ž.: *Zrcadlo alchymie*, Praha: Trigon, 1993.

Bor, D. Ž. et al.: *Opus Magnum*, Praha: Trigon, 1997.

Brettel, James A.: *God's other son*, San Jose: Writer's Showcase, 2000.

Buchwald, Jed Z. – Feingold, Mordechai: *Newton and the origin of civilization*, Princeton and Oxford: Princeton University Press, 2013.

Casini, Paolo: "Newton: The Classical Scholia," in: *History of Science*, XXII (1984), pp. 1–58.

Cotnoir, Brian: *Alchemy*, York Beach: Red Wheel/Weiser, LLC, 2006.

Cudworth, Ralph: *The True Intellectual System of the Universe, wherein all the reason and Philosophy of Atheism is confuted, and its Impossibility Demonstrated*, London, 1678.

De Smet, Rudolf – Verelst, Karin: "Newton's *Scholium Generale*: The Platonic and

Stoic legacy – Philo, Usus lipsius and the Cambridge Platonists," in: *History of Science*, XXXIX (2001), pp. 1–30.

Dobbs, Betty Jo Teeter: *The Foundations of Newton's Alchemy*, Cambridge: Cambridge University Press, 1975.

Dobbs, Betty Jo Teeter: *The Janus faces of genius*, Cambridge: Cambridge University Press, 2002.

Dobbs, Betty Jo Teeter: "Newton's Alchemy and His Theory of Matter," in: *Isis* LXXIII (1982), pp. 511–528.

Eliade, Mircea: *Dějiny náboženského myšlení III.*, Praha: Oikoymenh, 1997.

Eliade, Mircea: *Kováři a alchymisté*, Praha: Argo, 2000.

Everard, John: *The Divine Pymander*...http://www.levity.com/corpherm.html.

Faulkner, Raymond O.: *A Concise Dictionary of Middle Egyptian*, Oxford: Griffith Institute, 1988.

Festugiére, A.-J. (ed.): *Corpus hermeticum, Tome 2 – Traités 13–18, Asclepius*, Paris: Belles-Lettres, 1983.

Feynman, Richard P.: *O povaze fyzikálních zákonů*, Praha: Aurora, 2001.

Figala, Karin: "Newton as Alchemist," in: *History of Science* XV (1977), pp. 102–137.

Figala, Karin: "Die exakte Alchemie von Isaac Newton," in: *Verhandlungen der naturforschenden Gesellschaft in Basel*, Band 92, (1981), pp. 157–227.

Figala, Karin – Priesner, Claus (eds.): *Lexikon alchymie*, Praha: Vyšehrad, 2006.

Force, James E. – Popkin, Richard H. (eds.): *The books of nature and Scripture*, Dordrecht: Kluwer Academic 1994.

Force, James E. – Popkin, Richard H. (eds.): *Essays on the context, nature and influence of Isaac Newton's theology*, Dordrecht: Kluwer Academic, 1990.

Fulcanelli: *Příbytky filosofů*, Praha: Trigon, 1996.

Fulcanelli: *Tajemství katedrál*, Praha: Trigon, 1992.

Gebelein, Helmut: *Alchymie, magie hmoty*, Praha: Volvox Globator, 1998.

Haage, Bernard Ditrich: *Středověká alchymie*, Praha: Vyšehrad, 2001.

Hall, A. Rupert – Hall, Marie Boas: *Unpublished scientific papers of Isaac Newton*, Cambridge: Cambridge University Press, 1978.

Harrison, John: *The library of Isaac Newton,* Cambridge and New York: Cambridge University Press, 1978.

Hejdánek, Ladislav: *Filosofie a víra*, Praha: Oikoymenh, 1999.

Hoeller, Stephan A.: *C. G. Jung a gnóze*, Praha: Eminent, 2006.

Iliffe, Robert: *Persecution Complexes: The religious structure of Newton's philosophical conduct*, manuscript, 2004.

Iliffe, Robert: *Prisca Newtoniana*, manuscript, 2005.

Iliffe, Robert: *The snare of a beautiful hand and the unity of Newton's archive*, manuscript, 2006.

Jaspers, Karl: *Filosofická víra*, Praha: Oikoymenh, 1994.

Jaspers, Karl: *Šifry transcendence*, Praha: Vyšehrad, 2000.
Kovář, František: *Filosofické myšlení hellenistického židovstva*, Praha: Herrmann a synové, 1996.
Koyré, Alexander: *Od uzavřeného světa k nekonečnému vesmíru*, Praha: Vyšehrad, 2004.
Kozák, Jaromír: *Hermetismus. Tajné nauky starého Egypta*, Praha: Eminent, 2002.
Kratochvíl, Zdeněk: *Mýtus, filosofie, věda*, Praha: Hrnčířství a nakladatelství Michal Jůza & Eva Jůzová, 1993.
Kratochvíl, Zdeněk: *Prolínání světů,* Praha: Herrmann a synové, 1991.
Kuschi, Michio: *Kámen filosofů*, Bratislava: RI-EL/CAD Press, 1994.
Lasenic, Pierre de: *Alchymie, její teorie a praxe*, Praha: Půdorys, 1997.
Malíšek, Vladimír: *Isaac Newton – zakladatel teoretické fyziky*, Praha: Prometheus, 1999.
Mann, Thomas: *Joseph and His Brothers*, translated by John E. Woods, New York: Alfred A. Knopf, 2005.
Manuel, Frank E.: *The religion of Isaac Newton*, Oxford: The Claredon Press, 1974.
Manuel, Frank E.: *Isaac Newton, Historian*, Cambridge: The Belknap Press of Harvard University Press, 1963.
McGreal, Ian P. et al.: *Velké postavy západního myšlení*, Praha: Prostor, 1999.
McGuire, James E.: "The Fate of the Date: The Theology of Newton's *Principia* Revisited," in: *Rethinking the Scientific Revolution*, Margaret J. Osler (ed.), Cambridge: Cambridge University Press, 2000.
McGuire, J. E. – Rattansi, P. M.: "Newton and the 'Pipes of Pan'," in: *Notes and Records of the Royal Society of London*, Vol. 21, No. 2 (Dec. 1966), pp. 108–142.
Moran, Bruce T.: *Distilling knowledge. Alchemy, chemistry and the Scientific revolution*, Cambridge: Harvard University Press, 2005.
Nakonečný, Milan: *Lexikon magie*, Praha: Ivo Železný, 1994.
Nakonečný, Milan: *Smaragdová deska Herma Trismegista*, Praha: Vodnář, 1994.
Neubauer, Zdeněk: Apotheóza metamorfózy, in: *Akademie u sv. Mikuláše, Sborník přednášek 2004/2005*, Praha: Blahoslav, 2005.
Neubauer, Zdeněk: Corpus hermeticum Scientiae, *Logos, sborník pro esoterní chápání života a kultury*, Praha: Trigon, 1997.
Neubauer, Zdeněk: *Přímluvce postmoderny*, Praha: Hrnčířství a nakladatelství Michal Jůza & Eva Jůzová, 1994.
Newman, William R. – Principe, Lawrence, M.: *Alchemy tried in the Fire*, Chicago: The University of Chicago Press, 2002.
Newman, William R.: *Gehennical Fire, The lives of George Starkey*, Chicago: The University of Chicago Press, 2003.
Newton, Isaac: *Opticks, or a treatise of the reflections, refractions, inflections & colours of light*, 4th ed., 1730, New York: Dover, 1952.

Newton, Isaac: *Philosophiae naturalis principia mathematica*, Cantabrigiae, MDCCXIII.

Newton, Isaac: *The Principia: Mathematical principles of natural philosoph and his system of the world*, translated into English by Andrew Motte in 1729, revisited by Florian Cajori, Berkeley: University of California Press, 1974.

Newton, Isaac: *The Principia: Mathematical principles of natural philosophy*, a new translation by I. Bernard Cohen and Anne Whitman, assisted by Julia Budenz, Berkeley: University of California Press, 1999.

Nový, Luboš – Smolka, Josef: *Isaac Newton*, Praha: Orbis, 1969.

Otto, Rudolf: *Posvátno*, Praha: Vyšehrad, 1998.

Patricius, Franciscus: *Magia Philosophica, hoc est F. Patricii summi philosophi Zoroaster & eius 320 Oracula Chaldaica. Asclepp Dialogus. & Philosophia magna Hermetis Trismegisti* [...] *Latine reddita*. Hamburgi, 1593.

Platón: *Ústava*, Praha: Svoboda, 1993.

Popkin, Richard H.: "Newton's Biblical theology and his theological physics," in: *Newton's scientific and philosophical legacy*, (eds.) P. B. Scheuer – G. Debrock, Dordrecht: Kluiwer, 1988.

Prach, Václav: *Řecko-český slovník*, Praha: Springer et al., 1942.

Prigogine, Ilya – Stengersová, Isabelle: *Řád z chaosu,* Praha: Mladá fronta, 2001.

Principe, Lawrence M.: *The aspiring Adept. Robert Boyle and his alchemical Quest*. Princeton: Princeton University Press, 1998.

Rippe, Olaf et al.: *Paracelsovo lékařství*, Praha: Volvox Globator, 2004.

Röd, Wolfgang: *Novověká filosofie II*, Praha: Oikoymenh, 2004.

Roger, Bernard: *Objevování alchymie,* Praha: Malvern, 2005.

Rollo, Vlastimil: *Emocionalita a racionalita*, Praha: Sociologické nakladatelství (SLON), 1993.

Sailor, Danton B.: "Newton's debt to Cudworth," *Journal of the History of Ideas*, 49, (1988), pp. 511–518.

Saxl, Ivan: *Isaac Newton – alchymista, filosof, heretik*, manuscript, 2006.

Snobelen, Stephen D.: A time and times and the dividing of time, *Canadian Journal of History* 38 (2003), pp. 537–552.

Snobelen, Stephen D.: "God of Gods, and Lord of Lords: the theology of Isaac Newton's General Scholium to the Principia," in: *Osiris* 16, pp. 169–208.

Snobelen, Stephen D.: "Isaac Newton, heretic: the strategies of a Nicodemite," in: *The British Journal for the History of Science* 32 (1999) pp. 381–419.

Snobelen, Stephen D.: *Isaac Newton, Socianianism: Association with a Greater Heresy*, http://www.isaac-newton.org.

Snobelen, Stephen D.: *Lust, pride and ambition: Isaac Newton and the Devil*, http://www.isaac-newton.org.

Snobelen, Stephen D.: "The Mystery of the Restitution of All Things: Isaac Newton and the return of Jews," in: *Millenarianism and messianism in early Modern European Culture. The Millenarian Turn*, J. F. Force, R. H. Pipkin (eds.). Dordrecht: Kluwer Acad. Publ. 2001, pp. 95–118.

Snobelen, Stephen D.: *The Theology of Isaac Newton's General scholium to the Principia,* http://www.isaac-newton.org/.

Snobelen, Steven David: *To discourse of God: Isaac Newton's heterodox theology and his natural philosophy*, http://www.isaac-newton.org/.

Spisar, Alois: *Dějiny dogmatu*, Praha: Ústřední rada CČS, undated.

Struž, Jan – Studýnka, Bohumil: *Zlato – příběh neobyčejného kovu*, Praha: Grada, 2005.

Szydlo, Zbigniew: *Water which does not wet Hands. The Alchemy of Michael Sendivogius*. Warszawa: Polish Academy of Science, 1994.

Teilhard de Chardin, Pierre: *Vesmír a lidstvo*, Praha: Vyšehrad, 1993.

Thackray, Arnold: *Atoms and Powers. An Essay on Newtonian Matter-Theory and the Development of Chemistry*, Cambridge: Harvard University Press, 1970.

Tresmontant, Claude: *Bible a antická tradice*, Praha: Vyšehrad, 1998.

Tretera, Ivo: *Nástin dějin evropského myšlení*, Litomyšl: Paseka, 2002.

Trtík, Zdeněk: *Komparativní symbolika*, Praha: Husova československá bohoslovecká fakulta, 1962.

Trtík, Zdeněk: *Vztah já-ty a křesťanství*, Praha: Ústřední rada CČSH, 1948.

Vopěnka, Petr: *Geometrizace reálného světa (Třetí rozpravy s geometrií)*, Praha: Matfyzpress, 1995.

Wagner, Fritz: *Isaac Newton im Zwielicht zwischen Mythos und Forschung*, München: Verlag Karl Alber, 1976.

Westfall Richard: "Newton's Theological Manuscripts," in: *Contemporary Newtonian Research*, ed. Z. Bechler, Dordrecht: D. Reidel, 1982.

Westfall, Richard: "Newton and Christianity," in: *Facets of faith and science. Volume 3: The role of beliefs in the natural sciences. The Pascal Centre*, ed. J. M. van der Meer, Ancaster: The Pascal Centre, 1996.

Westfall, Richard: *The Life of Isaac Newton*, Cambridge: Cambridge University Press, 1999.

Yates, Frances A.: *Rozenkruciánské osvícenství*, Praha: Pragma, 2000.

Appendix 1

PHILOSOPHIAE NATURALIS PRINCIPIA MATHEMATICA AUCTORE ISAACO NEWTONO

Editio tertia MDCCXXVI
(Paginae ab 526 usque ad 530)
SCHOLIUM GENERALE

Hypothesis vorticum multis premitur difficultatibus. Ut planeta unusquisque radio ad solem ducto areas describat tempori proportionales, tempora periodica partium vorticis deberent esse in duplicata ratione distantiarum a sole. Ut periodica planetarum tempora sint in proportione sesquiplicata distantiarum a sole, tempora periodica partium vorticis-deberent esse in sesquiplicata distantiarum proportione. Ut vortices minores circum saturnum, jovem & alios planetas gyrati conserventur & tranquille natent in vortice solis, tempora periodica partium vorticis solaris deberent esse aqualia. Revolutiones solis & planetarum circum axes suos, qua cum motibus vorticum congruere deberent, ab omnibus hisce proportionibus discrepant. Motus cometarum sunt summe regulares, & easdem leges cum planetarum motibus observant, & per vortices explicari nequeunt. Feruntur cometa motibus valde eccentricis in omnes coelorum partes, quod fieri non potest, nisi vortices tollantur.

Projectilia, in aere nostro, solam aeris resistentiam sentiunt. Sublato aere, ut sit in vacuo Boyliano, resistentia cessat, siquidem pluma tenuis & aurum solidum aquali cum velocitate in hoc vacuo cadunt. Et par est ratio spatiorum coelestium, qua sunt supra atmospharam terra. Corpora omnia in istis spatiis liberrime moveri debent; & propterea planeta & cometa in orbibus specie & positione datis secundum leges supra expositas perpetuo revolvi. Perseverabunt quidem in orbibus suis per leges gravitatis, sed regularem orbium situm primitus acquirere per leges hasce minime potuerunt.

Planeta sex principales revolvuntur circum solem in circulis soli concentricis, eadem motus directione, in eodem plano quamproxime. Luna decem revolvuntur circum terram, jovem & saturnum in circulis concentricis, eadem motus directione, in planis orbium planetarum quamproxime. Et hi omnes mutus regulares originem non habent ex causis mechanicis; siquidem cometa in orbibus valde eccentricis, & in omnes coelorum partes libere feruntur. Quo motus genere cometa per orbes planetarum celerrime & facillime transeunt, & in apheliis suis ubi tardiores sunt & diutius morantur; quam longissime distant ab invicem, ut se mutuo quam minime trahant. Elegantissima hacce solis, planetarum & cometarum compages non nisi consilio & dominio entis intelligentis & potentis oriri potuit. Et si stella fixa sint centra similium systematum, hac omnia simili consilio constructa suberunt Unius dominio: prafertim cum lux fixarum sit ejusdem natura ac lux solis, & systemata omnia lucem in omnia invicem immittant. Et ne fixarum systemata per gravitatem suam in se mutuo cadant, hic eadem immensam ab invicem distantiam posuerit.

Hic omnia regit non ut anima mundi, sed ut universorum dominus. Et propter dominium suum, dominus deus Παντοκρατωρ[325] dici solet. Nam deus est vox relativa & ad servos refertur: & deitas est dominatio dei, non in corpus proprium, uti sentiunt quibus deus est anima mundi, sed in servos. Deus summus est ens aternum, infinitum, absolute perfectum: sed ens utcunque perfectum sine dominio non est dominus deus. Dicimus enim deus meus, deus vester, deus Israelis, deus deorum, & dominus dominorum: sed non dicimus aternus meus, aternus vester, aternus Isrealis, aternus deorum; non dicimus infinitus meus, vel perfectus meus. Ha appellationes relationem non habent ad servos. Vox deus passim[326] significat dominum: sed omnis dominus non est deus. Dominatio entis spiritualis deum constituit, vera verum, summa summum, ficta fictum. Et ex dominatione vera sequitur deum verum esse vivum, intelligentem & potentem; ex reliquis perfectionibus summum esse, vel summe perfectum. Aternus est & infinitus, omnipotens & omnisciens, id est, durat ab aterno in aternum, & adest ab infinitio in infinitum: omnia regit; & omnia cognoscit, qua fiunt aut fieri possunt. Non est aternitas & infinitas, sed aternus & infinitus; non est duratio & spatium, sed durat & adest. Durat semper, & adest ubique, & existendo semper & ubique, durationem & spatium constituit. Cum unaquaque spatii particula sit semper, & unumquodque durationis indivisibile momentum ubique, certe rerum omnium fabricator ac dominus non erit numquam, nusquam. Omnis

---

325 Id est Imperator universalis.

326 [*In margine: ll. 40ff.*] Pocockus noster vocem dei deducit a voce Arabica du, (& in casu obliquo di,) dua dominum significat. Et hoc sensu principes vocantur dii, Psalm lxxxiv. 6. & Joan. x. 45. Et Moses dicitur deus fratris Aaron, & deus regis Pharaoh (Exod. iv. 16. & vii 1.) Et eodem sensu anima principum mortuorum olim a gentibus vocabantur dii, sed falso propter defectum dominii.

anima sentiens diversis temporibus, & in diversis sensuum, & mortuum organis eadem est persona indivisibilis. Partes dantur successiva in duratione, coexistentes in spatio, neutra in persona hominis seu principio ejus cogitante; & multo minus in substantia cogitante dei. Omnis homo, quatenus res sentiens, est unus & idem homo durante vita sua in omnibus & singulis sensuum organis. Deus est unus & idem deus semper & ubique. Omniprasens est non per virtutem solam, sed etiam per substantiam: nam virtus sine substantia subsistere non potest. In ipso[327] continentur & moventur universa, sed sine mutua passione. Deus nihil patitur ex corporum motibus: illa nullam sentiunt resistentiam ex omniprasentia dei. Deum summum necessario existere in consesso est: Et eadem necessitate semper est & ubique. Unde etiam totus est sui similis, totus oculus, totus auris, totus cerebrum, totus brachium, totus vis sentiendi, intelligendi, & agendi, sed more minime humano, more minime corporeo, more nobis prorsus incognito. Ut cacus non habet ideam colorum, sic nos ideam non habemus modorum, quibus deus sapientissimus sentit & intelligit omnia. Corpore omni & figura corporea prorsus destituitur, ideoque videri non potest, nec audiri, nec tangi, nec sub specie rei alicujus corporei coli debet. Ideas habemus attributorum ejus, sed quid sit rei alicujus substantia minime cognoscimus. Videmus tantum corporum figuras & colores, audimus tantum sonos, tangimus tantum superficies externas, olfacimus odores solos, & gustamus sapores: intimas substantias nullo sensu, nulla actione reflexa cognoscimus; & multo minus ideam habemus substantia dei. Hunc cognoscimus solummodo per properietates ejus & attributa, & per sapientissimas & optimas rerum structuras & causas finales, & admiramur ob perfectiones; veneramur autem & colimus ob dominium. Colimus enim ut servi, & deus sine dominio, providentia, & causis finalibus nihil aliud est quam fatum & natura. A caca necessitate metaphysica, qua utique eadem est semper & ubique, nulla oritur rerum variatio. Tota rerum conditarum pro locis ac temporibus diversitas, ab ideis & voluntate entis necessario existentis solummodo oriri potuit. Dicitur autem deus per allegoriam videre, audire, loqui, ridere, amare, odio habere, cupere, dare, accipere, gaudere, irasci, pugnare, fabricare,

327 [*In margine: ll. 57ff.*] Ita sentiebant veteres, ut Pythagoras apud Ciceronem, de Natura deorum, lib. 1.

Thales, Anaxagoras, Virgilius Georgic. lib iv. v. 220, & Aneid. lib 6. v. 721. Phile Allegor. lib. 1. sub initio. Aratus in Phanom. sub initio. Ita etiam scriptores sacri ut Paulus in Act. xvii. 27, 28. Johannes in Evang. xiv. 2. Moses in Deut. iv. 39. & x. 14. David Psal. cxxxix. 7, 8, 9. Solomon 1 Reg. viii. 27 Jeremias xxiii. 23, 24. Fingebant autem idololatra solem, lunam, & astra, animas hominium & alias mundi partes esse partes dei summi & ideo colendas sed falso.

condere, construere. Nam sermo omnis de deo a rebus humanis per similitudinem aliquam desumitur, non perfectam quidem, sed aliqualem tamen. Et hac de deo, de quo utique ex phanomenis disserere, ad philosophiam naturalem pertinet.

Hactenus phanomena calorum & maris nostri per vim gravitatis exposui, sed causam gravitatis nonum assignavi. Oriur utique hac vis a causa aliqua, qua penetrat ad usque centra solis & planetarum, sine virtutis diminutione; quaque agit non pro quantitate superficierum particularum, in quas agit (ut solent causa mechanica) sed pro quantitate materia solida; & cujus actio in immensas distantias undique extenditur, decrescendo semper in duplicata ratione distantiarum. Gravitas in solem componitur ex gravitatibus in singulas solis particulas, & recedendo a sole decrescit accurate in duplicata ratione distantiarum ad usque orbem saturni, ut ex quiete apheliorum planetarum manifestum est, & ad usque ultima cometarum aphelia, si modo aphelia illa quiescant. Rationem vero harum gravitatis proprietatum ex phanomenis nondum potui deducere, & hypotheses non fingo. Quicquid enim ex phanomenis non deducitur, hypothesis vocanda est; & hypotheses seu metaphysica, seu physica, seu qualitatum occultarum, seu mechanica, in philosophia experimentali locum non habent. In hac philosophia propositiones deducuntur ex phanomenis, & redduntur generales per inductionem. Sic impenetrabilitas, mobilitas, & impetus corporum & leges motuum & gravitatis innotuerunt. Et satis est quod gravitas revera existat, & agat secundum leges a nobis expositas, & ad corporum calestium & maris nostri motus omnes sufficiat.

Adjicere jam liceret nonnulla de spiritu quodam subtilissimo corpora crassa pervadente, & in iisdem latente; cujus vi & actionibus particula corporum ad minimas distantias se mutuo attrahunt, & contigua facta coharent; & corpora electrica agunt ad distantias majores, tam repellendo quam attrahendo corpuscula vicina; & lux emittitur, reflectitur, refringitur, inflectitur, & corpora calefacit; & sensatio omnis excitatur, & membra animalium ad voluntatem moventur, vibrationibus scilicet hujus spiritus per solida nevrorum capillamenta ab externis sensuum organis ad cerebrum & a cerebro in musculos propagatis. Sed hac paucis exponi non possunt; neque adest sufficiens copia experimentorum, quibus leges actionum hujus spiritus accurate determinari & monstrari debent.

FINIS

Appendix 2

The General Scholium to Isaac Newton's *Principia mathematica*

Published for the first time as an appendix to the 2nd (1713) edition of the Principia, *the General Scholium reappeared in the 3rd (1726) edition with some amendments and additions. As well as countering the natural philosophy of Leibniz and the Cartesians, the General Scholium contains an excursion into natural theology and theology proper. In this short text, Newton articulates the design argument (which he fervently believed was furthered by the contents of his* Principia*), but also includes an oblique argument for a unitarian conception of God and an implicit attack on the doctrine of the Trinity, which Newton saw as a post-Biblical corruption. The English translation here is that of Andrew Motte (1729). Italics and orthography as in original.*[1]

General Scholium:

The hypotheses of Vortices is press'd with many difficulties. That every Planet by a radius drawn to the Sun may describe areas proportional to the times of description, the periodic times of the several parts of the Vortices should observe the duplicate proportion of their distances from the Sun. But that the periodic times of the Planets may obtain the sesquiplicate proportion of their distances from the Sun, the periodic times of the parts of the Vortex ought to be in sesquiplicate proportion of their distances. That the smaller Vortices may maintain their lesser revolutions about *Saturn*, *Jupiter*, and other Planets, and swim quietly and undisturb'd in the greater Vortex of the Sun, the periodic times of the parts of the Sun's Vortex should be equal. But the rotation of the Sun and Planets about their axes, which ought to correspond with the motions of their Vortices, recede far from all these proportions. The motions of the Comets are exceedingly regular, are govern'd by the same laws with the motions of the Planets, and can by no means be accounted for by the hypotheses of Vortices. For Comets are carry'd with very eccentric motions through all parts of the heavens indifferently, with a freedom that is incompatible with the notion of a Vortex.

Bodies, projected in our air, suffer no resistance but from the air. Withdraw the air, as is done in Mr. *Boyle*'s vacuum, and the resistance ceases. For in this void a bit of fine down and a piece of solid gold descend with equal velocity. And the parity of reason must take place in the celestial spaces above the Earth's atmosphere; in which spaces, where there is no air to resist their motions, all bodies will move with the greatest freedom; and the Planets and Comets will constantly pursue their revolutions in orbits given in kind and position, according to the laws above explain'd. But though these bodies may indeed persevere in their orbits by the mere laws of gravity, yet they could by no means have at first deriv'd the regular position of the orbits themselves from those laws.

The six primary Planets are revolv'd about the Sun, in circles concentric with the Sun, and with motions directed towards the same parts and almost in the same plan. Ten Moons are revolv'd about the Earth, Jupiter and Saturn, in circles concentric with them, with the same direction of motion, and nearly in the planes of the orbits of those Planets. But it is not to be conceived that mere mechanical causes could give birth to so many regular motions: since the Comets range over all parts of the heavens, in very eccentric orbits. For by that kind of motion they pass easily through the orbits of the Planets, and with great rapidity; and in their aphelions, where they move the slowest, and are detain'd the longest, they recede to the greatest distances from each other, and thence suffer the least disturbance from their mutual attractions.

This most beautiful System of the Sun, Planets, and Comets, could only proceed from the counsel and dominion of an intelligent and powerful being. And if the fixed Stars are the centers of other like systems, these, being form'd by the like wise counsel, must be all subject to the dominion of One; especially since the light of the fixed Stars is of the same nature with the light of the Sun, and from every system light passes into all the other systems. And lest the systems of the fixed Stars should, by their gravity, fall on each other mutually, he hath placed those Systems at immense distances from one another.

This Being governs all things, not as the soul of the world, but as Lord over all: And on account of his dominion he is wont to be called *Lord God Pantokrator*[2], or *Universal Ruler*. For *God* is a relative word, and has a respect to servants; and Deity is the dominion of God, not over his own body, as those imagine who fancy God to be the soul of the world, but over servants. The supreme God is a Being eternal, infinite, absolutely perfect; but a being, however perfect, without dominion, cannot be said to be Lord God; for we say, my God, your God, the God of *Israel*, the God of Gods, and Lord of Lords; but we do not say, my Eternal, your Eternal, the Eternal of *Israel*, the Eternal of Gods; we do not say, my Infinite, or my Perfect: These are titles which have no respect to servants. The word *God* usually a [3] signifies *Lord*; but every lord is not a God. It is the dominion of a spiritual being which constitutes a God; a true, supreme, or imaginary dominion makes a true, supreme, or imaginary God. And from his true dominion it follows that the true God is a Living, Intelligent, and Powerful Being; and, from his other perfections, that he is Supreme or most Perfect. He is Eternal and Infinite, Omnipotent and Omniscient; that is, his duration reaches from Eternity to Eternity; his presence from Infinity to Infinity; he governs all things, and knows all things that are or can be done. He is not Eternity and Infinity, but Eternal and Infinite; he is not Duration and Space, but he endures and is present. He endures forever, and is every where present; and, by existing always and every where, he constitutes Duration and Space. Since every particle of Space is *always*, and every indivisible moment of Duration is *every where*, certainly the Maker and Lord of all things cannot be *never* and *no where*. Every soul that has perception is, though in different times and in different organs of sense and motion, still the same indivisible person. There are given successive parts in duration, co-existent parts in space, but neither the one nor the other in the person of a man, or his thinking principle; and much less can they be found in the thinking substance of God. Every man, so far as he is a thing that has perception, is one and the same man during his whole life, in all and each of his organs of sense. God is the same God, always and everywhere. He is omnipresent, not *virtually* only, but also *substantially*; for virtue cannot subsist without substance. In him b [3] are all things contained and moved; yet neither affects the other: God suffers nothing from the motion of bodies; bodies find no resistance from the omnipresence of God. 'Tis allowed by all that the supreme God exists necessarily; and by the same necessity he exists *always* and *every where*. Whence also he is all similar, all eye, all ear, all brain, all arm, all power to perceive, to understand, and to act; but in a manner not at all human, in a manner not at all corporeal, in a manner utterly unknown to us. As a blind man has no idea of colours, so have we no idea of the manner by which the all-wise God perceives and understands all things. He is utterly void of all body and bodily figure, and can therefore neither be seen, nor heard, not touched; nor ought he to be worshipped under the representation of any corporeal thing. We have ideas of his attributes, but what the real substance of anything is we know not. In bodies, we see only their figures and colours, we hear only the sounds, we touch only their outward surfaces, we smell only the smells, and taste the savours; but their inward substances are not to be known, either by our senses, or by any reflex act of our minds; much less then have we any idea of the substance of God. We know him only by his most wise and excellent contrivances of things, and final causes; we admire him for his perfections; but we reverence and adore him on account of his dominion. For we adore him as his servants; and a God without dominion, providence, and final causes, is nothing else but Fate and Nature. Blind metaphysical necessity, which is certainly the same always and every where, could produce no variety of things. All that diversity of natural things which we find, suited to different times and places, could arise from nothing but the ideas and will of a Being necessarily existing. But, by way of allegory, God is said to see, to speak, to laugh, to love, to hate, to desire, to give, to receive, to rejoice, to be angry, to fight, to frame, to work, to build. For all our notions of God are taken from the ways of mankind, by a certain similitude which, though not perfect, has some likeness, however. And thus much concerning God; to discourse of whom from the appearances of things, does certainly belong to Natural Philosophy. [5]

Hitherto we have explain'd the phænomena of the heavens and of our sea, by the power of Gravity, but have not yet assign'd the cause of this power. This is certain, that it must proceed from a cause that penetrates to the very centers of the Sun and Planets, without suffering the least diminution of its force; that operates, not according to the quantity of surfaces of the particles upon which it acts, (as mechanical causes use to do,) but according to the quantity of the solid matter which they contain, and propagates its virtue on all sides, to immense distances, decreasing always in the duplicate proportion of the distances. Gravitation towards the Sun, is made up out of the gravitations towards the several particles of which the body of the Sun is compos'd; and in receding from the Sun, decreases accurately in the duplicate proportion of the distances, as far as the orb of Saturn, as evidently appears from the quiescence of the aphelions of the Planets; nay, and even to the remotest aphelions of the Comets, if those aphelions are also quiescent. But hitherto I have not been able to discover the cause of those properties of gravity from phænomena, and I frame no hypotheses. For whatever is not deduc'd from the phænomena, is to be called an hypothesis; and hypotheses, whether metaphysical or physical, whether of occult qualities or mechanical, have no place in experimental philosophy. In this philosophy particular propositions are inferr'd from the phænomena, and afterwards render'd general by induction. Thus it was that the impenetrability, the mobility, and the impulsive force of bodies, and the laws of motion and of gravitation, were discovered. And to us it is enough, that gravity does really exist, and act according to the laws which we have explained, and abundantly serves to account for all the motions of the celestial bodies, and of our sea.

And now we might add something concerning a certain most subtle Spirit, which pervades and lies hid in all gross bodies; by the force and action of which Spirit, the particles of bodies mutually attract one another at near distances, and cohere, if contiguous; and electric bodies operate to greater distances, as well repelling as attracting the neighbouring corpuscles; and light is emitted, reflected, refracted, inflected, and heats bodies; and all sensation is excited, and the members of animal bodies move at the command of the will, namely, by the vibrations of this Spirit, mutually propagated along the solid filaments of the nerves, from the outward organs of sense to the brain, and from the brain into the muscles. But these are things that cannot be explain'd in few words, nor are we furnish'd with that sufficiency of experiments which is required to an accurate determination and demonstration of the laws by which this electric and elastic spirit operates.

Notes

[1] Isaac Newton, *The Mathematical principles of natural philosophy*, trans. Andrew Motte (London, 1729), pp. 387–93.

[2] *Pantokrator*: original in Greek.

[3] Newton's note a: Dr. *Pocock* derives the Latin word *Deus* from the *Arabic du* (in the oblique case *di*,) which signifies *Lord*. And in this sense Princes are called *Gods*, *Psal.* lxxxii. ver. 6; and *John* x. ver. 35. And *Moses* is called a *God* to his brother *Aaron*, and a *God* to *Pharaoh* (*Exod.* iv. ver. 16; and vii. ver. 1 [correction for the 1729 edition, which reads: 8]). And in the same sense the souls of dead princes were formerly, by the Heathens, called *gods*, but falsely, because of their want of dominion. [This note was added to the 3rd, 1726 edition].

[4] Newton's note b: This was the opinion of the Ancients. So *Pythagoras* in *Cicer. de Nat. Deor.* lib. i. *Thales*, *Anaxagoras*, *Virgil*, Georg. lib. iv. ver. 220. and Aeneid. lib. vi. ver. 721. *Philo Allegor.* at the beginning of lib. i. *Aratus* in his Phænom. at

the beginning. So also the sacred Writers, as St. *Paul*, Acts xvii. ver. 27, 28. St. *John*'s Gosp. chap. xiv. ver. 2. *Moses* in *Deut*. iv. ver. 39; and x. ver. 14. *David*, *Psal*. cxxxix. ver. 7, 8, 9. *Solomon*, 1 *Kings* viii. ver. 27. *Job* xxii. ver. 12, 13, 14. *Jeremiah* xxiii. ver. 23, 24. The Idolaters supposed the Sun, Moon, and Stars, the Souls of Men, and other parts of the world, to be parts of the supreme God, and therefore to be worshiped; but erroneously.

[5] 1713 edition: Experimental Philosophy.

Appendix 3

Quotes from Corpus Hermeticum, used by Ralph Cudworth in T.I.S. in order as they appeared there, and from other Treatises, used by Zdeněk Kratochvíl (in parentheses at the end is the indication of Treatise).

(1a) **Nonne hoc dixi, Omnia unum esse, et unum omnia, utpote quia in creatore fuerint omnia, antiquam creasset omnia? Nec immerito unus est dictus omnia, cuius membra sunt omnia...** Have we not already declared that all things are one, and one all things? Forasmuch as all things existed in the Creator, before they were made; neither is he improperly said to be all things, whose members all things are. (Asclepius)

(1b) **Huius itaque, qui est unus omnia, vel ipse est Creator omnium, in tota hac disputatione curato meminisse.** Be thou therefore mindful in this whole disputation of him, who is one and all things, or was the creator of all. (Asclepius)

(2) **Idcirco non erant, quando nata non erant, sed in eo jam tunc erant, unde nasci habuerunt ...** And yet at that very time were they in him, from whom they were afterwards produced... (Asclepius)

(3) **Non spero totius majestatis effectorem, omnium rerum patrem vel dominium, uno posse quamvis e multis composito nomine nuncupari. Hunc voca potius omni nomine, siquidem sit unus et omnia; ut necesse sit aut omnia ipsius nomine, aut ipsum omnium nomine nuncupari.** I cannot hope sufficiently express the author of majesty, and the father and lord of all things, by any one name, though compounded of never so many names. Call him therefore by every name, forasmuch as he is one and all things; so that of necessity. Either all things must be called by his name, or he by the names of all things. (Asclepius)

(4a) **Solus deus ipse in se, et a se, et circum se, totus est plenus atque perfectus** ... God alone in himself, and from himself, and about himself, is altogether perfect. (Asclepius)

(4b) ... **isque sua firma stabilitas est; nec alicujus impulsu, nec loco moveri potest, cum in eo sint omnia, et in omnibus ipse est solus** ... and himself is his own stability. Neither can he be moved or changed by the impulse of any thing, since all things are in him, and he alone is in all things... (Asclepius)

(5a) **Hic sensibilis mundus receptaculum est omnium sensibilium specierum, qualitatum, vel corporum** ... The sensible world is the receptacle of all forms, qualities and bodies... (Asclepius)

(5b) **... quae omnia sine Deo vegetari non possunt: Omnia enim Deus, et a Deo omnia et sine hoc, nec fuit aliquid, nec est, nec erit; omnia enim ab eo, et in ipso, et per ipsum ...** all which cannot be vegetated and quickened without God, for God is all things, and all things are from God, and all things the effect of his will; and without God there neither was any thing, nor is, nor shall be; but all things are from him, and in him, and by him... (Asclepius)

(5c) **Si totum animadvertes, vera ratione perdisces, mundum ipsum sensibilem, et quae in eo sunt omnia, a superiore illo mundo, quasi vestimento, esse contecta ...** And if you will consider things after a right manner, you shall learn that this sensible world, and all the things herein, are covered all over with that superior world (or deity) as it were with a garment. (Asclepius)

(6) The divinity is the whole mundane compages, or constitution; for nature is also placed in the deity. (III)

(7) For there is nothing in the whole world which he is not, he is both the things that are, and the things that are not; for the things that are he hath manifested; but the things that are not, he contains within himself... (V)

(8) ... he is all things that are, and therefore he hath all names, because all things are from one fater and therefore he hath no name, because he is the father of all things ... He is both incorporeal and omnicorporeal, for there is nothing of any body which he is not. (V)

(9) And for what cause shall I praise thee? Because I am my own, as having something proper and distinct from you? Thou are whatsoever I am; thou art whatsoever I do, or say, for thou art all things, and there is

nothing which thou art not; thou are that which is made, and thou art that which is unmade ... For what shall I praise thee? For those things which thou hast made, or for those things thou hast hidden and concealed within thyself? (V)

(10) Understand that the whole world is from God, and in God; for God is the beginning,
comprehension and constitution of all things. (VIII)

(11) I would not say that God hath all things, but rather declare the truth, and say he is all things; not as receiving them from without, but as sending them forth from himself. (IX)

(12) There shall never be a time, when any thing that is shall cease to be; for when I say any thing that is, I say any thing of God; for God hath all things in him, and there is neither any thing without God, nor God without any thing. (IX)

(13) What is God, but the very being of all things that yet are not, and the subsistence of things that are? (X)

(14) God is both the father and good, because he is all things. (X)

(15) God acting immediately from himself is always in his own work, himself being that which he makes; for if that were never so little separated from him, all would of necessity fall to nothing and die. (XI)

(16) All things are in God. But not as lying in a place. (XI)

(17) ... you may consider God in the same manner, as containing the whole world within himself, as his own conceptions and cogitations... (XI)

(18) Is God invisible? Speak worthily of him, for who is more manifest than he? For this very reason did he make all things, that you mightest see him through all things: this is the virtue and goodness of the Deity, to be seen through all things. The mind is seen in thinking, but God in working or making. (XI)

(19) I have heard the good daemon, (for he alone, as the first begotten god, beholding all things, spake divine words), I have heard him sometimes saying, that one is all things. (XII)

(20) This whole world is intimately united to him, and observing the order and will of its father, hath the fullness of life in it; and there is nothing in it through eternity which does not live; for there neither is, nor hath been, nor shall be, any thing dead in the world. (XII)

(21) This is God, the universe or all. And in this universe there is nothing which he is not: wherefore there is neither magnitude, nor place, nor quality, nor figure, nor time about God, for he is all or the whole. (XII)

(22a) I am about to praise the Lord of the creation... (XIII)

(22b) All the powers, that are in me, praise the one and the all. (XIII)

(23) If any one go about to separate the all from one, he will destroy the all, or the universe, for all ought to be one. (XV)

(24) I will begin with a prayer to him, who is the Lord, and maker, and father, and bound of all things ... is one, and being one, is all things; for the fulness of all things is one and in one. (XVI)

(25) And all things are parts of God, but if all things be parts of God, then God is all things; wherefore he making all things, doth, as it were, make himself. (XVI)

Irena Štěpánová

Newton

Kosmos – Bios – Logos

Published by Charles University in Prague
Karolinum Press
Ovocný trh 3/5, 116 36 Prague 1
Prague 2014
Vice-Rector-Editor Prof. PhDr. Ivan Jakubec, CSc.
Edited by Alena Jirsová
Language correction by Martina Pranic
Layout by Jan Šerých
Typeset by DTP Karolinum
Printed by Karolinum Press
First English Edition

ISBN 978-80-246-2379-5
ISBN 978-80-246-2389-4 (pdf)